Contents

Discovering map skills

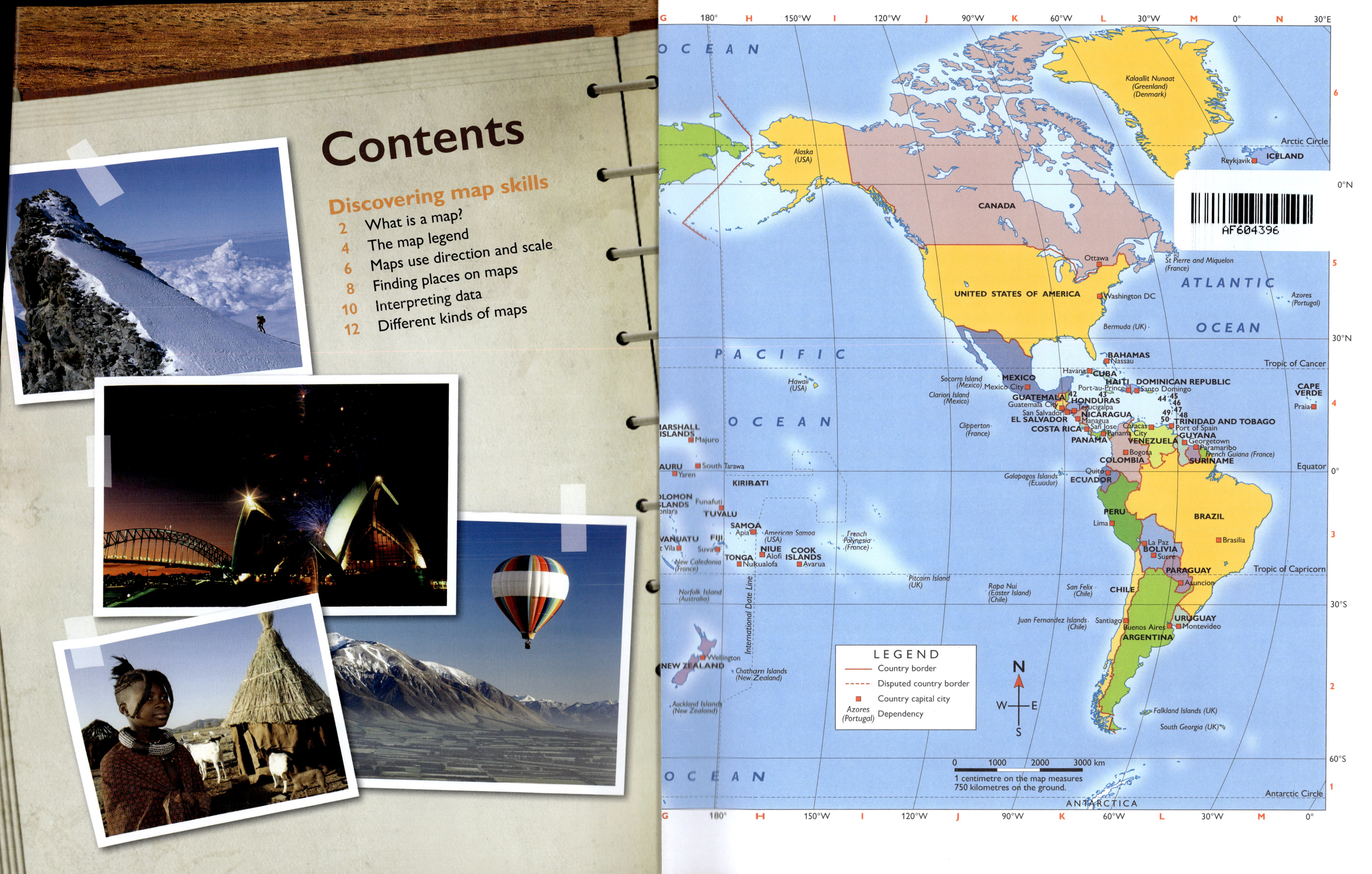

AF604396

OXFORD ATLAS FOR AUSTRALIAN SCHOOLS

3–4

HASS | STEM | Inquiry | Coding

OXFORD UNIVERSITY PRESS
AUSTRALIA & NEW ZEALAND

Discovering our world

Earth, Sun and Moon

Changing Earth

Natural resources

Living things

Heat

Forces

First people

Arrival and impact

Communities and celebrations

Discovering our country

Discovering continents and countries

DISCOVERING MAP SKILLS

What is a map?

Photographs from above

This photograph is taken from an aeroplane. It shows what the area looks like from above at one point in time. You can see the roofs of the buildings, basketball courts, roads and even cars on the roads.

Aerial photograph of Eltham North Primary School

Maps

Maps are also viewed from directly above, but maps are different from photographs. Maps use colours and symbols to show different features. A legend tells you what the colours and symbols mean. Labels tell you the names of the streets and important buildings. Only permanent features are shown. Cars and pedestrians are not shown on maps.

discover!

Look at the aerial photograph of Eltham North Primary School opposite.

Draw a tree from directly above.

Draw a tree from ground level.

What features on the aerial photograph don't appear on the map below? Why?

Map of Eltham North Primary School

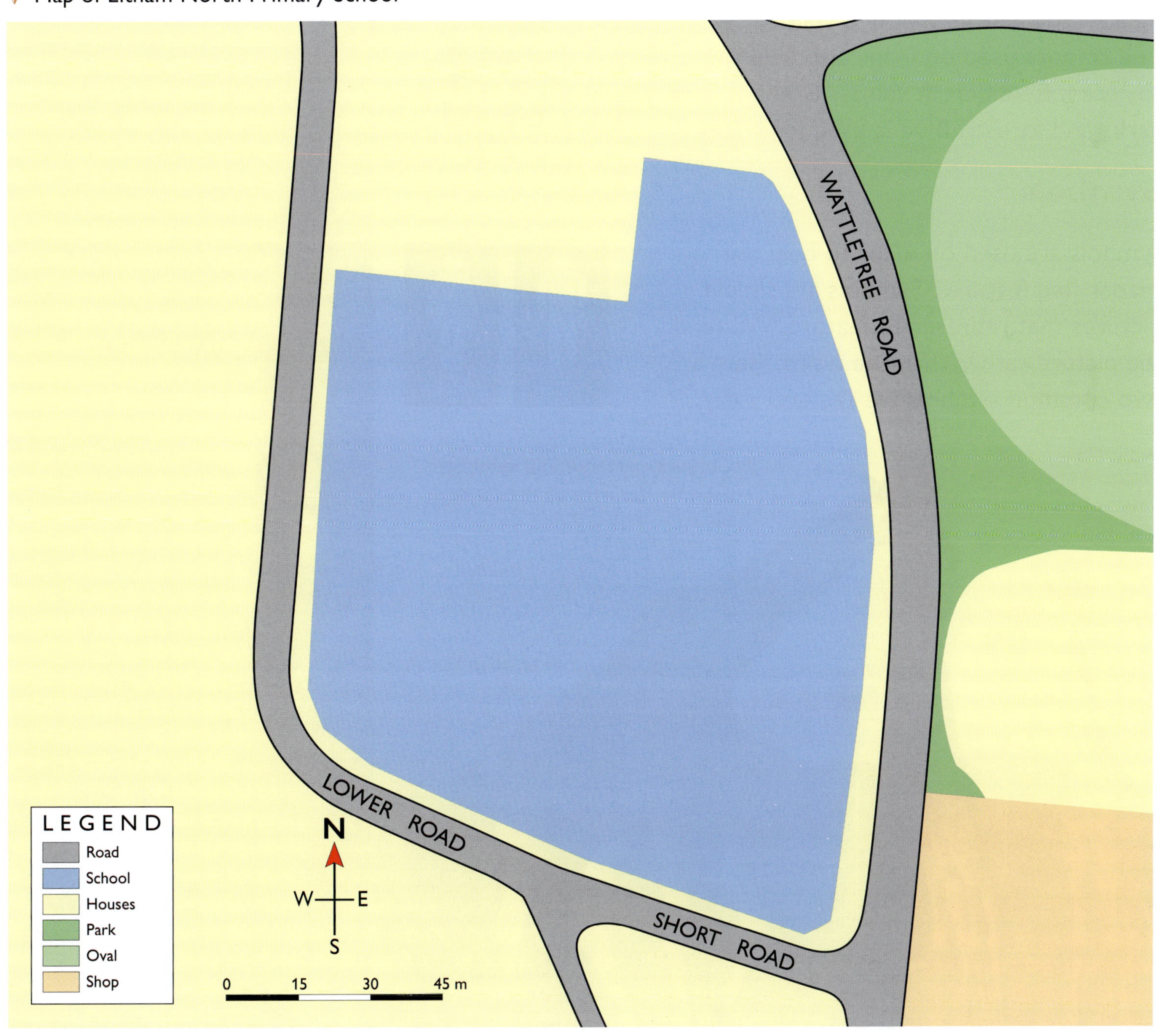

The map legend

The map legend explains the colours, patterns and symbols used on the map.

Colours

We know what the colours on a traffic light mean—red is stop, yellow is wait and green is go. The colours used on maps also help the reader to recognise different things. Green is used for parks and forests. Blue is used for water.

Symbols

Symbols are used on maps to help the reader find features. Symbols are simple pictures that represent a feature. We read the picture rather than the word. Here are two common symbols we see every day.

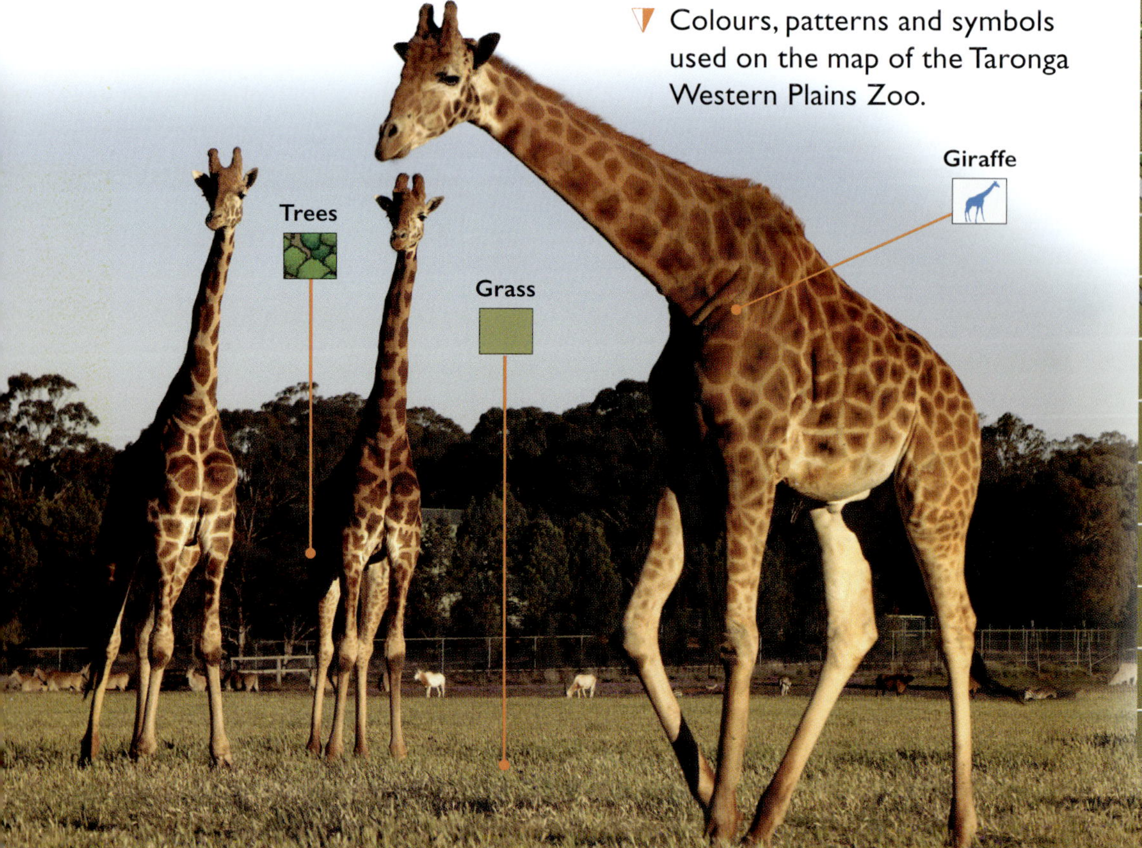

Colours, patterns and symbols used on the map of the Taronga Western Plains Zoo.

discover!

What animal can be found on an island in Savannah Lake?

Go to the red '4 km' label on the map. Name three Australian animals found near here. (Not all of these animals are listed in the legend.)

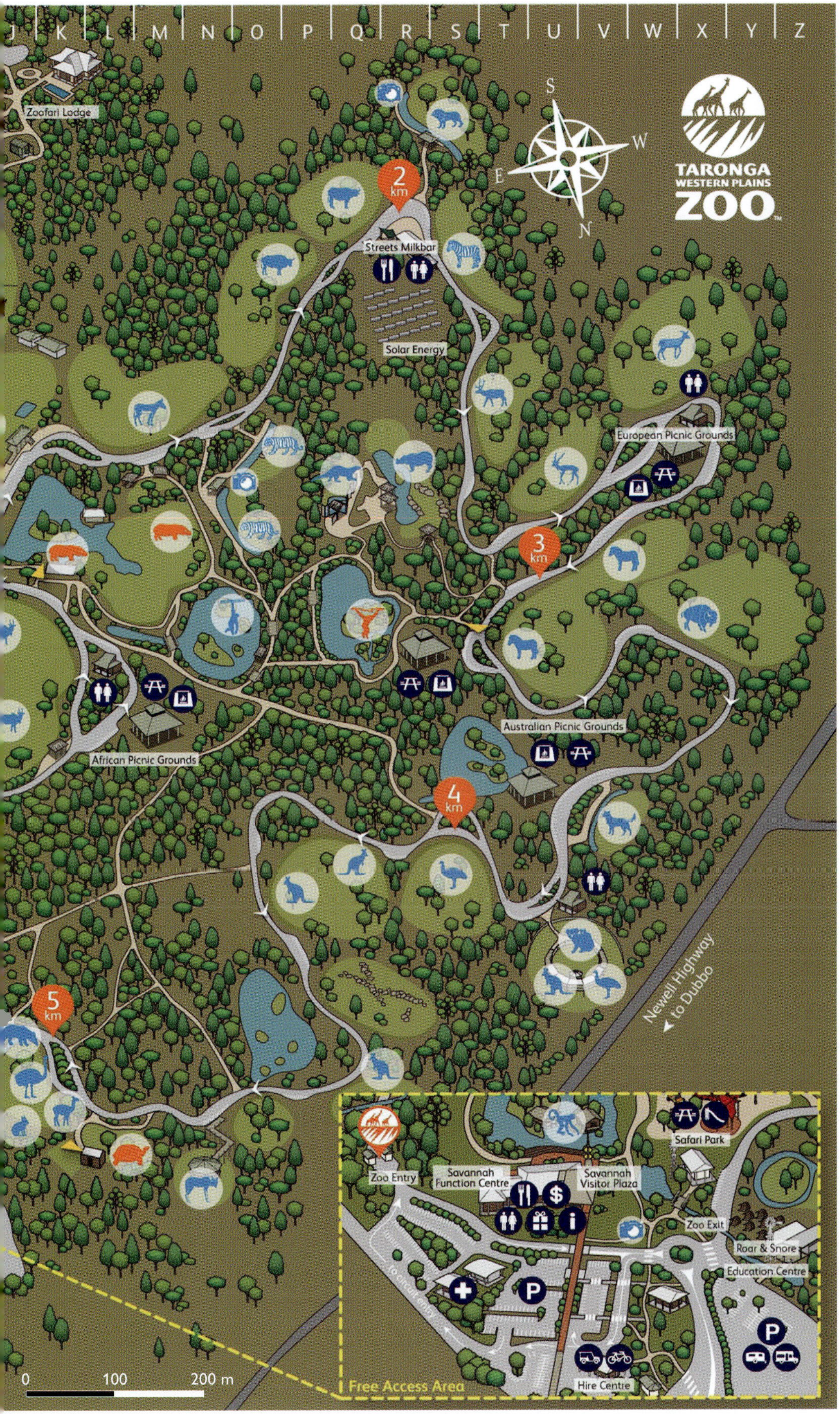

◄ Map of Taronga Western Plains Zoo

Maps use direction and scale

Describing the location of places

We use the points of a compass to give directions and to work out the location of features. Compass points always remain the same, no matter what direction you are facing.

Western Australia, South Australia and the Northern Territory all describe their location using compass points. Queensland, New South Wales and Victoria are all located on the east coast of Australia.

Measuring distances on maps

We can use the scale on the map to measure real distances.

1. Use the edge of a piece of paper to mark the distance between Melbourne and Sydney.
2. Hold the marks against the line scale at the bottom of the map, with one mark on zero.
3. Estimate the distance between Melbourne and Sydney.
4. Is Adelaide or Sydney closer to Melbourne?

◄ Measuring straight distances with a scale

States and territories

discover!

What direction is located between north and east?

What do you notice about the names of the animals on the map?

Turn to page 102. Use the line scale to measure the distance between Bangkok (B4) and Phnom Penh (B4).

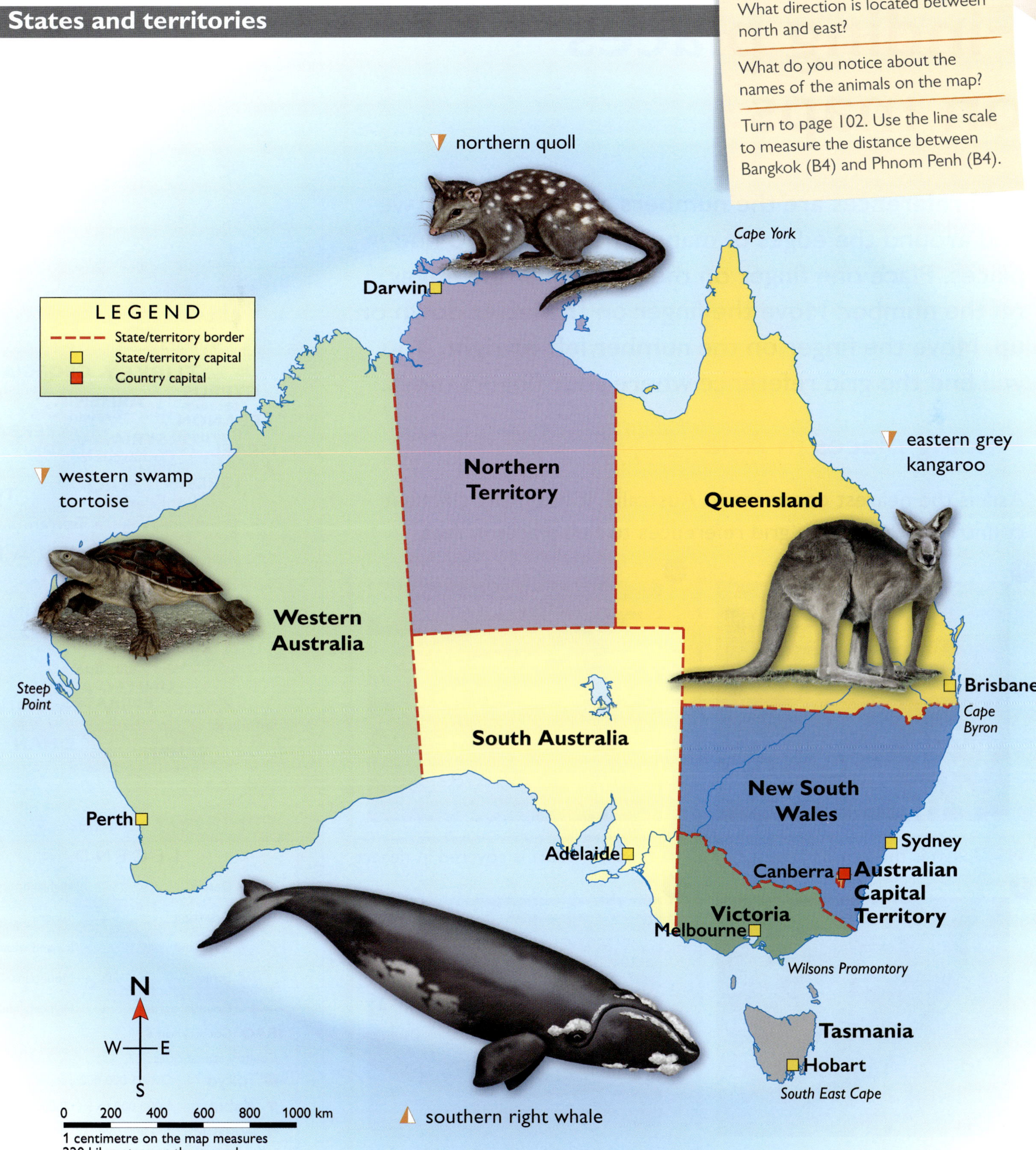

Finding places on maps

Grid references are the numbers and letters that we find around the edges of maps. They help us to find places. Place one finger on the letter and one finger on the number. Move the finger on the letter down or up. Move the finger on the number left or right. You will find the grid reference where your fingers meet.

Finding places in Asia

Asia is the nearest continent to Australia. It has many different countries. We can use grid references to find places in Asia.

1

H5 Tokyo is the world's largest city.

2

A4 Saudi Arabia is covered in desert.

3

G2 Orang-utans live in the forests of Malaysia.

4

D4 Mount Everest is the world's highest mountain.

discover!

What capital city is found at D2?

Manila is the capital city of the Philippines. What is its grid reference?

What capital city is found at B2 on page 94?

Interpreting data

Maps, tables and graphs can give readers a lot of information. They help you see patterns and trends in the data presented. These pages display different information about bilbies in the Arid Recovery Reserve in South Australia. See what you can learn about these bilbies by carefully studying the table, graphs and map.

Bilbies have been released into the Arid Recovery Reserve in outback Australia. The reserve is fenced to keep out predators such as feral cats, rabbits and foxes.

Data tables

Data tables arrange information into a series of rows and columns. This makes it easier to compare or tally information.

Bilbies released into the Arid Recovery Reserve

Year	Males	Females	Total
2000	4	5	9
2003	4	4	8
2004	11	4	15
2005	5	5	10

Column graphs

Column graphs show information in a column or a bar. They help us to easily compare things or to see trends.

Bilby population in the Arid Recovery Reserve

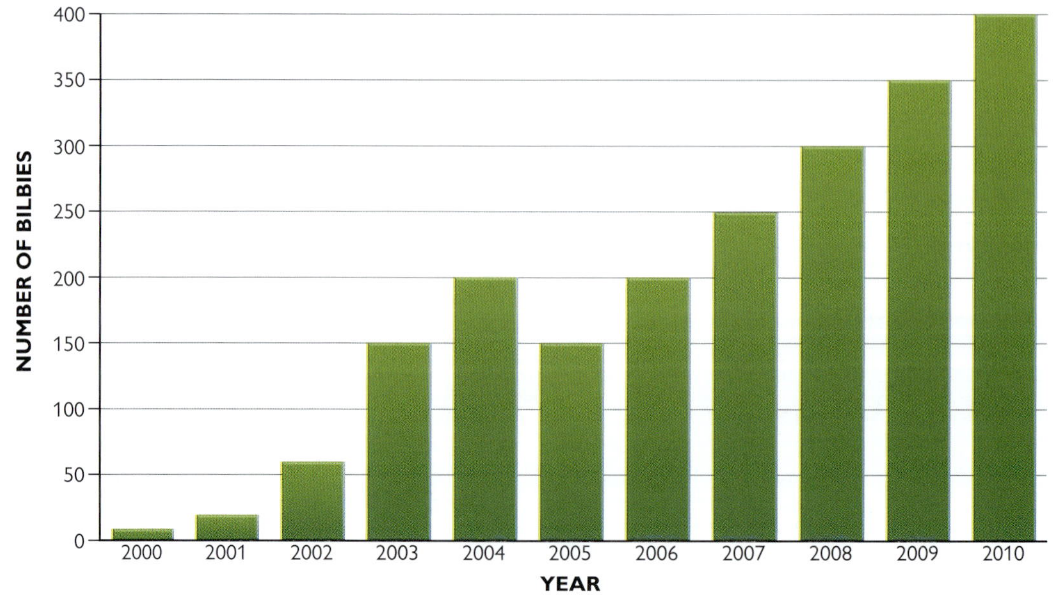

Picture graphs

Picture graphs use symbols to represent different amounts. This helps the reader to understand the data. Often symbols on picture graphs can represent more than one.

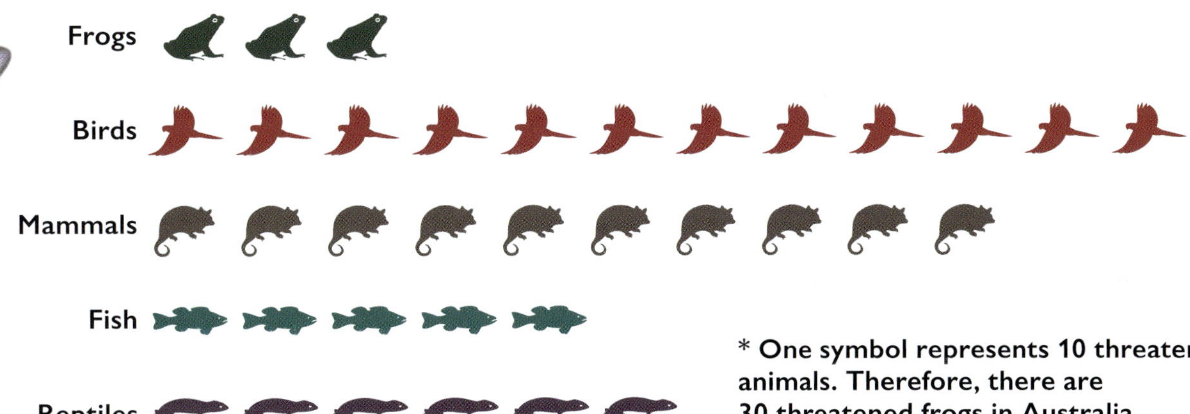

discover!

In what year were 15 bilbies released into the Arid Recovery Reserve?

How many threatened mammals are represented on the picture graph?

Look at the graph on page 108. How tall is the Eiffel Tower?

Maps

Data can also be shown on maps to let the reader know where something occurs. Maps can hold a lot of information:

- the names and borders of countries and states
- the locations of towns, cities and places
- distribution patterns—in this case, the areas where bilbies are found in Australia.

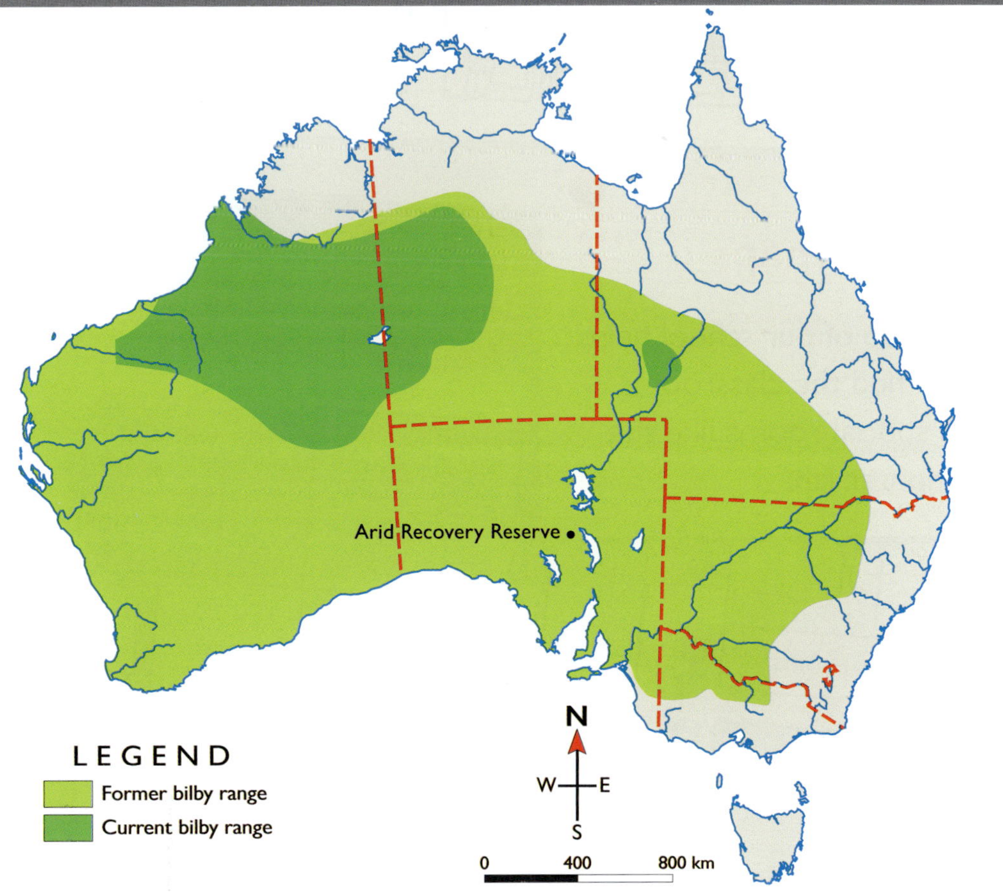

Different kinds of maps

You can see a range of maps every day. Maps are presented in different ways, depending on the type of information that is needed.

My map

We can draw our own maps to show friends where we live and how to get there.

In the car

Many cars have satellite navigation systems where a voice will guide you to your destination.

School maps

The map of our school helps us to find the classrooms and special rooms like the computer lab.

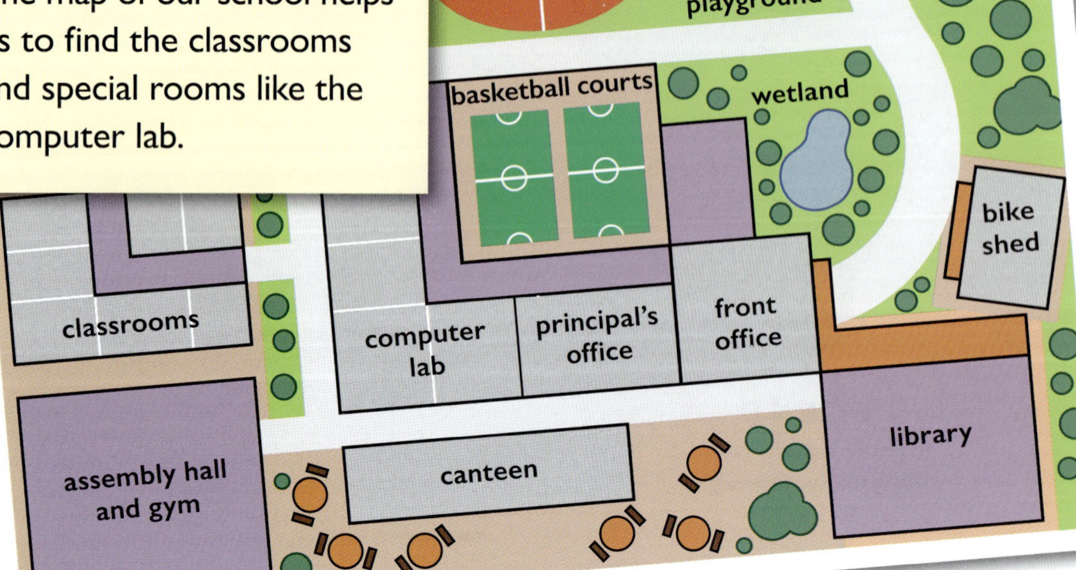

At the shops

Maps at shopping centres help you find your favourite store.

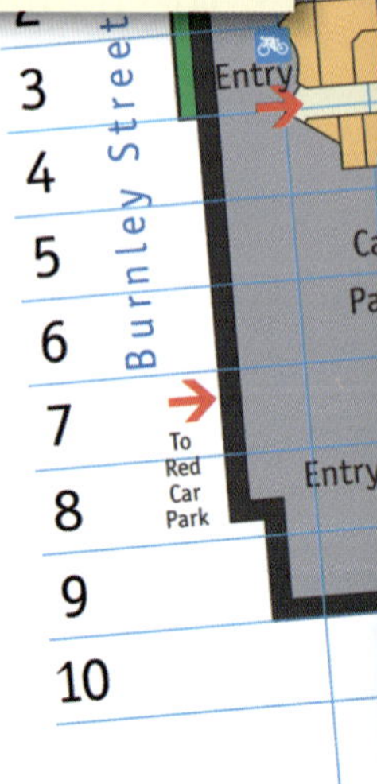

discover!

Draw a map of your classroom.

What do the 'H' and 'L' stand for on the weather map below?

What do the lines on the map on page 57 show?

In my atlas

An atlas has lots of different maps. To find a place, go to the index at the back of the atlas. Places are listed in alphabetical order to make them easy to find.

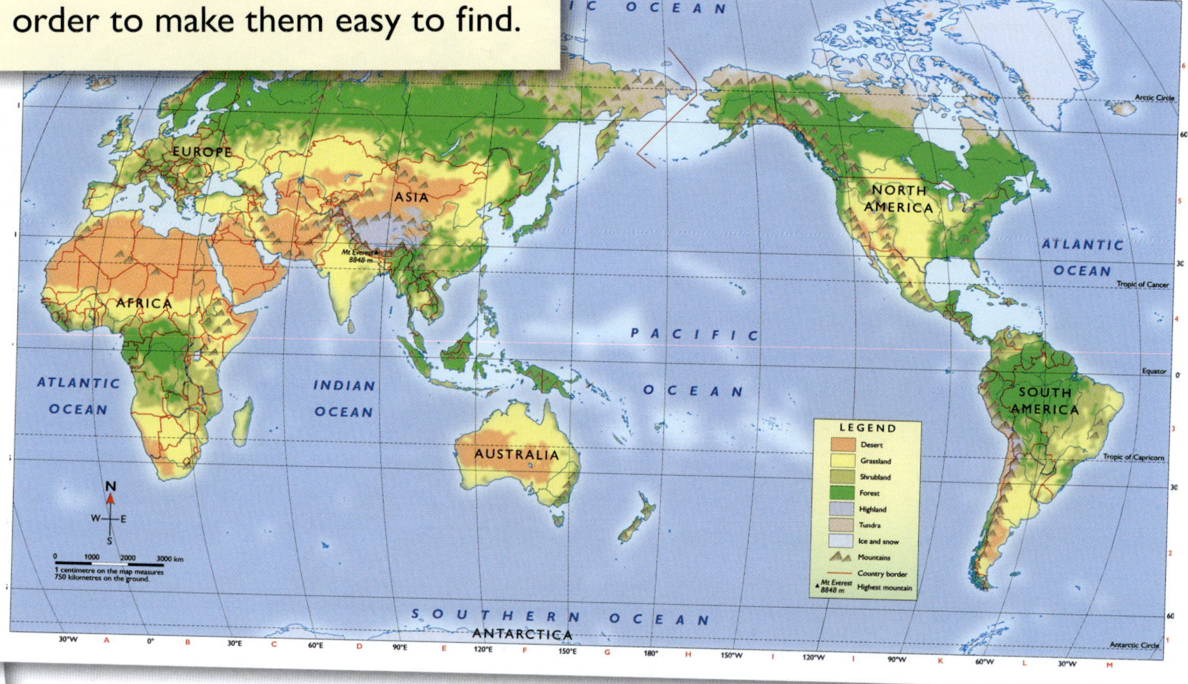

On television

Each night on the news we see a weather map. The 'H' stands for high pressure when the weather is fine. The 'L' stands for low pressure, which may bring rain.

DISCOVERING
OUR WORLD

Earth, Sun and Moon

Earth is the only planet in our solar system that supports life. The Sun provides light and warmth for life on Earth. The Sun always lights up one side of the Moon so that we can see it from Earth.

Sun

The Sun is a star. It is a huge ball of hot gas. It is the closest star to Earth and the only star visible during the day. The Sun is 110 times bigger than the Earth.

The temperature at the centre of the Sun is thought to be 14 million degrees Celsius.

discover!

How long does it take the Earth to orbit around the Sun?

What impact does the Moon have on Earth? (page 125 ▶▶)

What is aurora australis? (page 139 ▶▶)

Earth

The Earth is a planet. Planets are smaller objects that orbit around stars. It takes the Earth 365 days (one year) to orbit around the Sun.

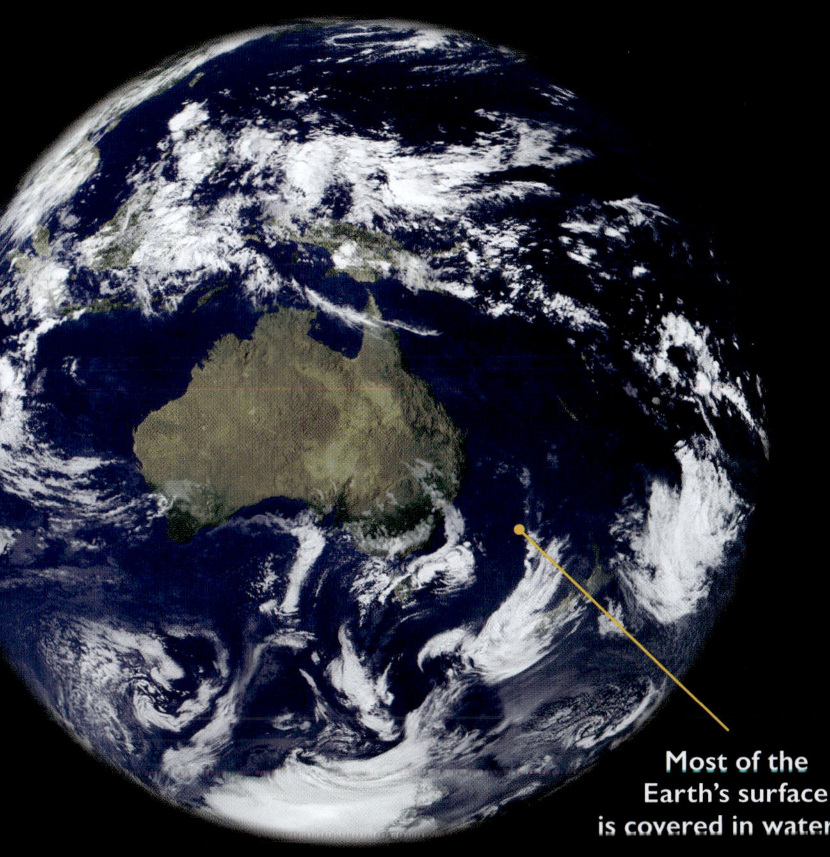

Most of the Earth's surface is covered in water.

Moon

Moons are objects that orbit planets. Our Moon takes 28 days to orbit the Earth. The Earth is four times bigger than the Moon.

Phases of the Moon

The Moon does not make its own light. Each night we see the part of the Moon facing the Sun. As the Moon orbits the Earth, we see it at different angles. These are called 'phases' of the Moon.

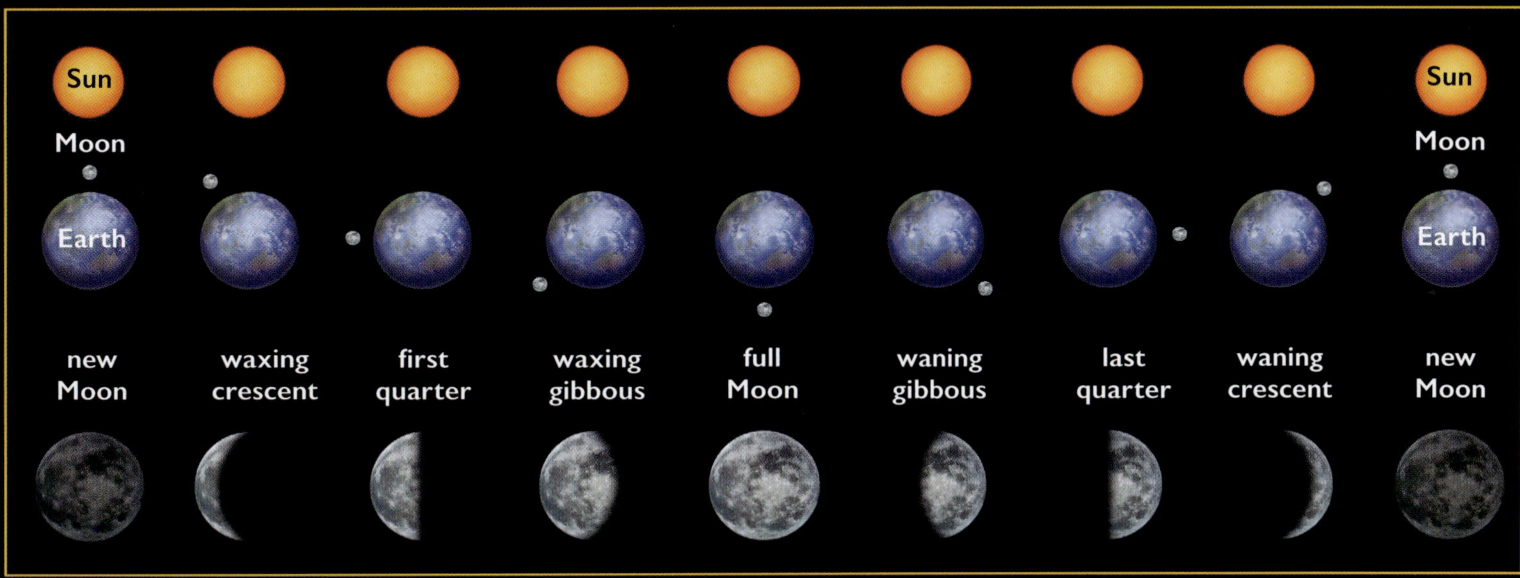

▲ Phases of the Moon seen in Australia

Day and night

Day and night are caused by the Earth rotating once every 24 hours. The part of Earth that faces the Sun is having day, while the part of Earth that faces away from the Sun is having night. People see the Sun rise in the east when the Earth spins towards the Sun. People see the Sun set in the west when the Earth spins away from the Sun.

The Sun rises over the horizon in the east.

The side of the Earth facing the Sun experiences day.

The Earth rotates 15 degrees every hour. It rotates a full circle (360 degrees) every day.

The Earth rotates towards the east.

Sunset is happening at the same time for all places along this line.

9 am
8 am
7 am
6 am
5 am
4 am
10 am
11 am
12 noon
1 pm
2 pm
3 pm
4 pm
5 pm
6 pm
7 pm
8 pm
9 pm
10 pm
Prime Meridian
Date Line
ASIA
PACIFIC OCEAN
AUSTRALIA

The side of the Earth facing away from the sun experiences night.

2 am

am

dnight

discover!

How long does it take Earth to rotate fully?

When is your shadow at its longest?

Sunlight and shadows

Objects cast a shadow on the opposite side to the Sun. In the middle of the day, the Sun is overhead and your shadow is short. Your shadow is long early in the morning and late in the afternoon.

This boy's shadow is short as it is close to midday.

This girl's shadow is long as it is early in the morning.

Sundials

A sundial can be used to tell the time. It has a pointer that casts a shadow. When the Sun follows its path from east to west across the sky, the shadow moves to show the time.

This sundial shows the time is about 10 o'clock in the morning.

Seasons

Each year, the Earth takes 365 days to orbit the Sun. During this orbit, the amount of sunlight reaching Australia changes, causing our seasons. When it is summer in Australia (in the southern hemisphere) it is winter in the northern hemisphere.

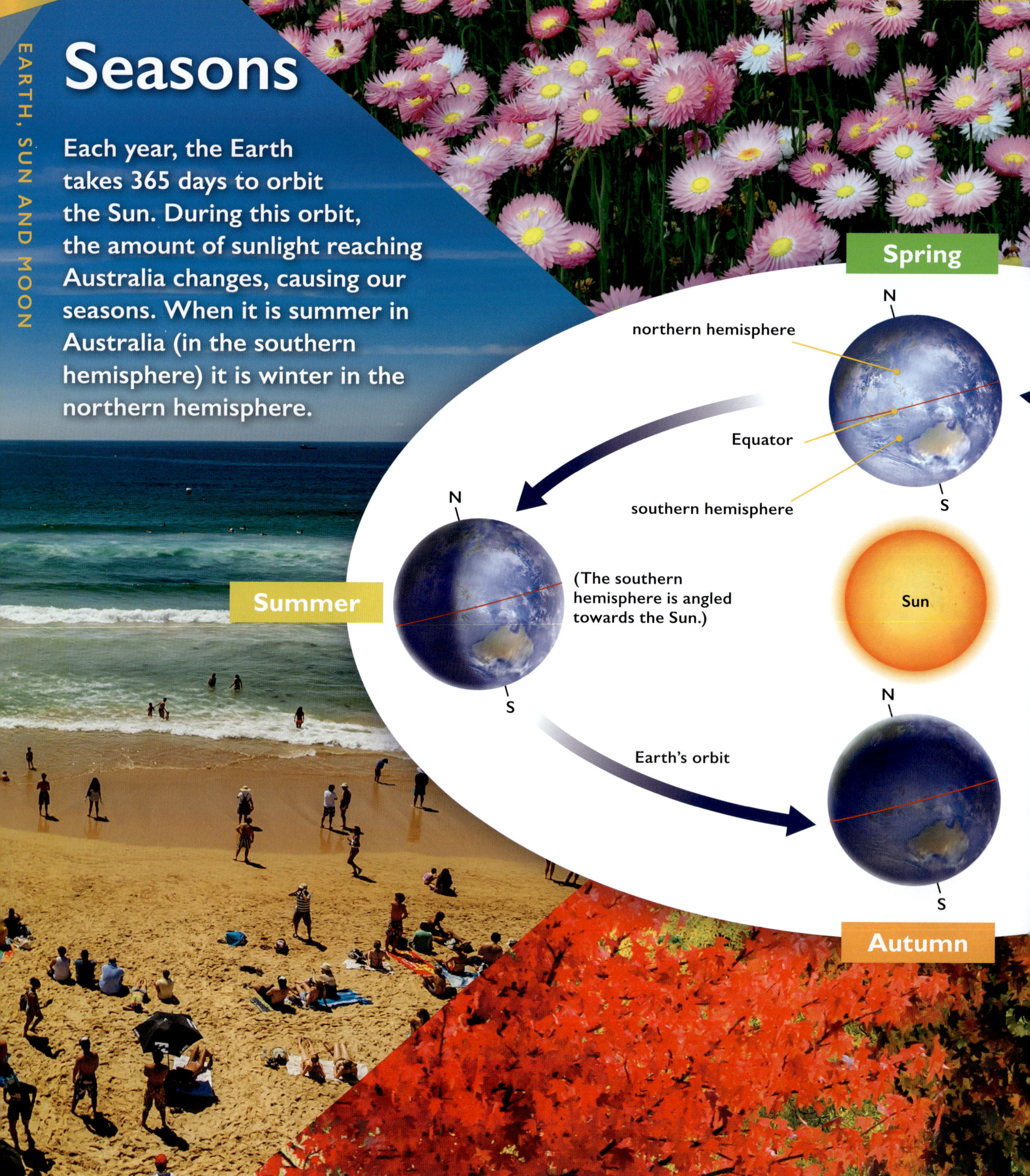

discover!

What season is it in Australia in January?

Northern Australia has only two seasons. What are they? (page 76 ▶▶)

How do changing seasons affect transport along canals in Amsterdam? (page 111 ▶▶)

Months

Months	Season	Description
December, January, February	**Summer**	The days are hottest and longest.
March, April, May	**Autumn**	The days become cooler and some leaves change colour.
June, July, August	**Winter**	The days are coldest and shortest.
September, October, November	**Spring**	The days become warmer and flowers bloom.

Rivers change the land

Rivers begin in mountains and flow down towards the sea. The running water wears away the land in a process called erosion. As rivers flow down from the mountains, they cut deep valleys or gorges. When they reach flatter ground, rivers form big loops called meanders.

The Katherine River in the Northern Territory has cut a deep gorge.

Floods

When there is a lot of rain, water builds up in rivers. Sometimes rivers cannot hold all of the water. Water then spills over the river bank and onto the land causing a flood.

Major floods can cover roads and wash into houses. During a flood, people may need to be rescued by boats. It may also be necessary to move farm animals to higher ground.

In 2013, the town of Bundaberg, in Queensland, experienced its worst flooding on record. The Burnett River broke its banks and flooded the town.

discover!

Where does the Ganges River begin? (page 106 ▶▶)

What has erosion by the Colorado River formed? (page 124 ▶▶)

What has the Amazon River formed as it flows towards the sea? (page 132 ▶▶)

The changing coastline

As waves crash into the coast they create landforms with many different shapes. The waves erode the land to form steep cliffs and caves. Sometimes a cave can become so large that it forms an arch. The rock that breaks away is smashed into smaller pieces by the waves. The smallest rock pieces form the sand that we find on beaches.

The coast near Port Campbell in Victoria is made of a soft rock called limestone. The waves have eroded the rocks to form shapes like this arch.

Headlands are steep, high parts of the coast that extend out into the sea. The area between two headlands is called a bay.

discover!

What is the main cause of erosion along the coast?

How does a cave become an arch?

Where are the caves in Halong Bay? (page 102 ▶▶)

As the rocks are worn away by waves, the grains of rocks gather on the beach as sand.

wave

Groynes are walls built on the beach to stop the sand washing away.

The roots of plants on sand dunes hold the sand together.

bay

sea

Headlands are made from hard rocks that are left behind when the softer rock has worn away.

stack

arch

cave

cliff

Humans change the land

Human activities change the surface of the land. We clear forests to build roads, farms and towns, as well as for logging and mining. Clearing the Earth's forests to use the land for a different purpose is called deforestation. Trees protect the land from erosion. When trees are removed, rain washes the topsoil away. The habitats of plants, animals and indigenous peoples are also destroyed.

Risk of deforestation

EUROPE
ASIA
North Kore
AFRICA
Nigeria
Cambodia
Democratic Republic of Congo
ATLANTIC OCEAN
Indonesia
Papu New Guine
INDIAN OCEAN
AUSTRALIA
N
W E
S
0 2000 4000 km

The Amazon rainforest

Trees are cut down and burned.

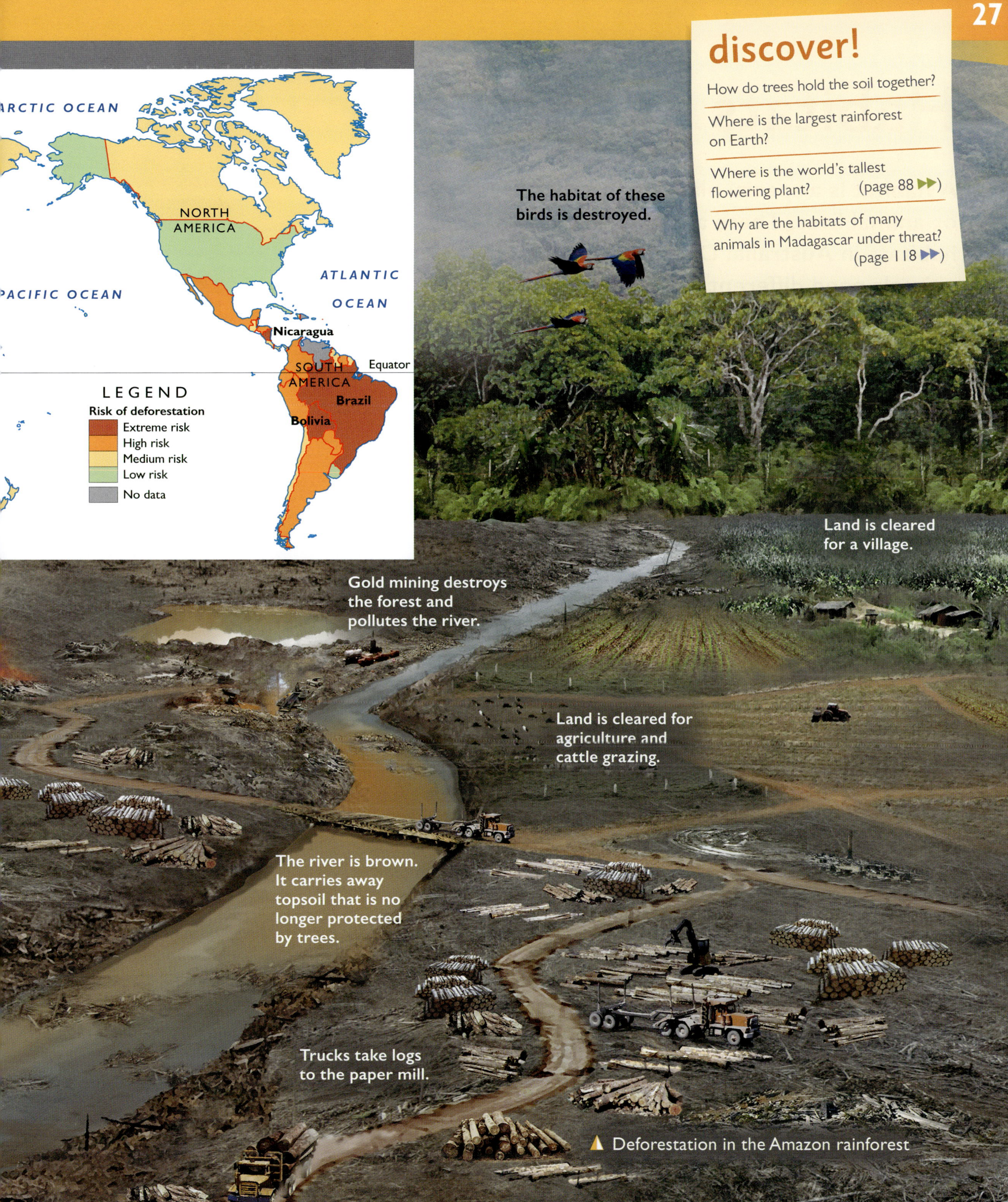

discover!

How do trees hold the soil together?

Where is the largest rainforest on Earth?

Where is the world's tallest flowering plant? (page 88 ▶▶)

Why are the habitats of many animals in Madagascar under threat? (page 118 ▶▶)

▲ Deforestation in the Amazon rainforest

Habitats in Australia

A habitat is a special place where a plant or animal lives. There are many kinds of habitats in Australia. Different plants and animals need different habitats—some live in the dry desert while others live in forests. Habitats provide animals with water, food, shelter and a safe place to raise their young. When a habitat is destroyed, the survival of some plants and animals can be threatened.

A kangaroo's habitat

Kangaroos don't need much water. They can adapt to most habitats, although they generally live in grasslands and on the edge of deserts. Kangaroos are consumers (eaters). They mostly eat grass (producer). The main predators (hunters) of kangaroos are dingoes and humans.

sunlight
grass
(producer)
kangaroo
(consumer)
human
(predator)
dingo
(predator)

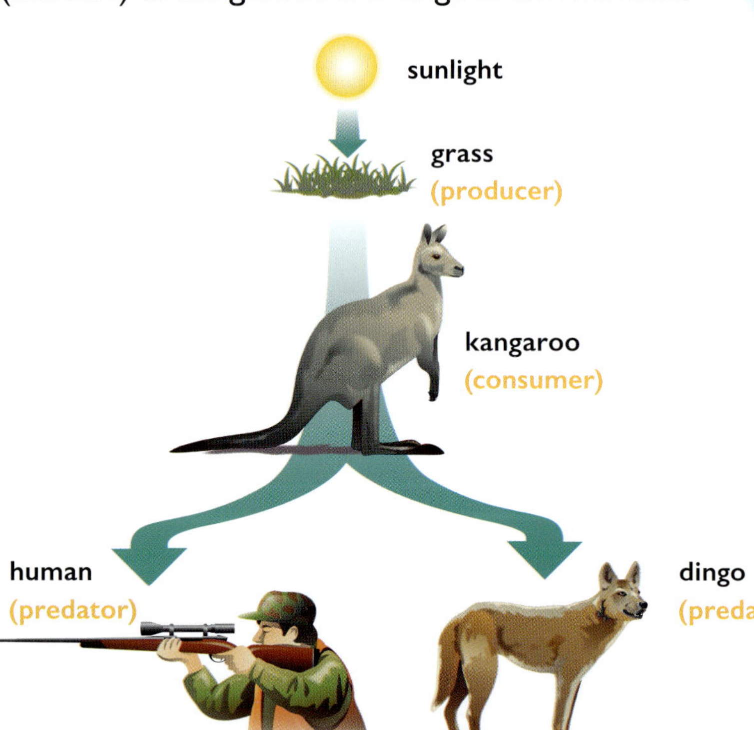

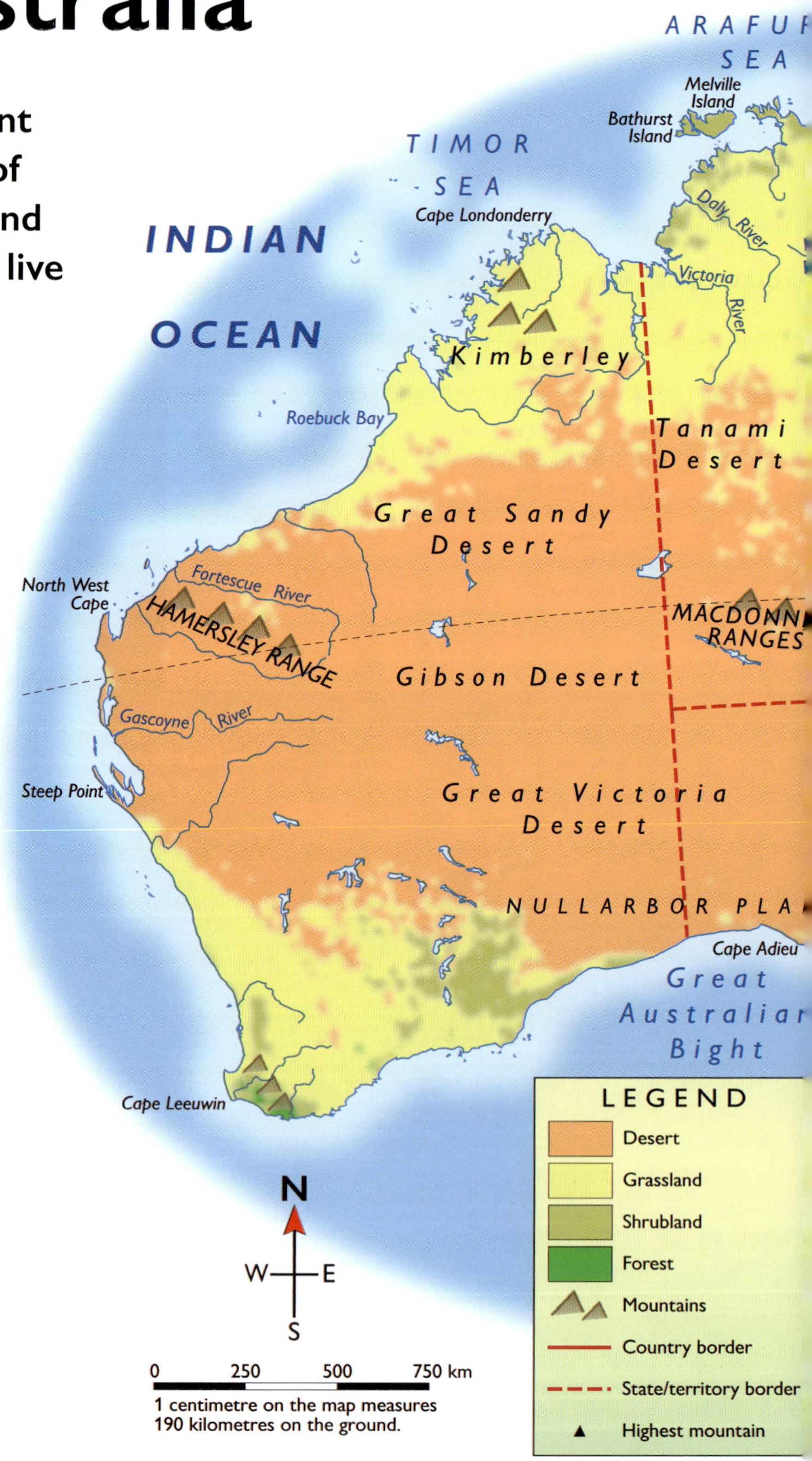

ARAFURA SEA
Melville Island
Bathurst Island
TIMOR SEA
Cape Londonderry
INDIAN OCEAN
Daly River
Victoria River
Kimberley
Roebuck Bay
Tanami Desert
Great Sandy Desert
North West Cape
Fortescue River
HAMERSLEY RANGE
MACDONNELL RANGES
Gibson Desert
Gascoyne River
Steep Point
Great Victoria Desert
NULLARBOR PLAIN
Cape Adieu
Great Australian Bight
Cape Leeuwin
N
W
E
S
0
250
500
750 km
1 centimetre on the map measures 190 kilometres on the ground.
LEGEND
Desert
Grassland
Shrubland
Forest
Mountains
Country border
State/territory border
Highest mountain

discover!

What habitats are shown on the map?

What is the habitat of the chimpanzee? (page 120 ▶▶)

Why is the giant panda's habitat under threat? (page 101 ▶▶)

A mushroom's habitat

This small, red mushroom's habitat is the forest. Mushrooms are not plants, they are fungi. Fungi are decomposers that break down dead plants and animals into nutrients. These nutrients are returned to the soil to be used by plants (producers).

Changing habitats

Habitats can change due to humans altering the land, or from natural disasters such as bushfires or floods. Humans digging channels for boats have caused the destruction of seagrass beds in Queensland. These seagrass beds can also be covered by soil washed down rivers during floods. With no seagrass (producer), consumers such as turtles and dugongs can die.

Australia's natural resources

Natural resources are things from the natural environment that we use. Sunlight, wind, water and forests are natural resources. Other natural resources are oil, coal, natural gas and minerals. Australia has some of the world's largest mineral and energy deposits. Natural resources need to be managed carefully so they are available for future generations.

Forests are used for bushwalking.

Forests are used for logging.

Forests are habitats for animals.

Wind is used to produce electricity.

Coal is mined from the Earth.

Minerals and energy

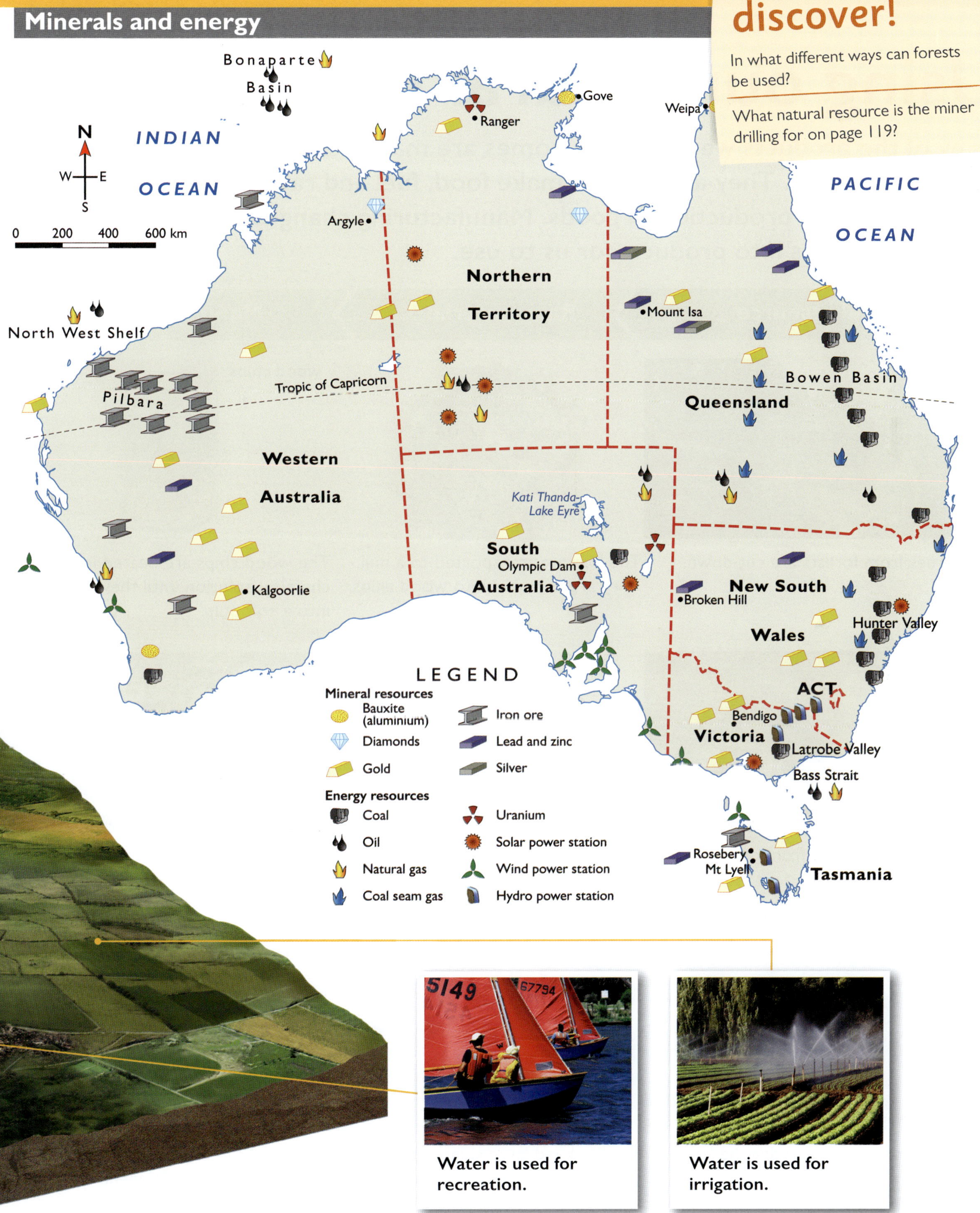

Water is used for recreation.

Water is used for irrigation.

discover!

In what different ways can forests be used?

What natural resource is the miner drilling for on page 119?

Using our natural resources

Many of the products we use in our homes are made from natural resources. They are used to make food, fuel and raw materials for the production of goods. Manufacturing changes natural resources into products for us to use.

Natural resource	Raw material	Manufacturing

PAPER

Trees from forests are cut down.

The logs are transported to a mill. They are chopped into wood chips.

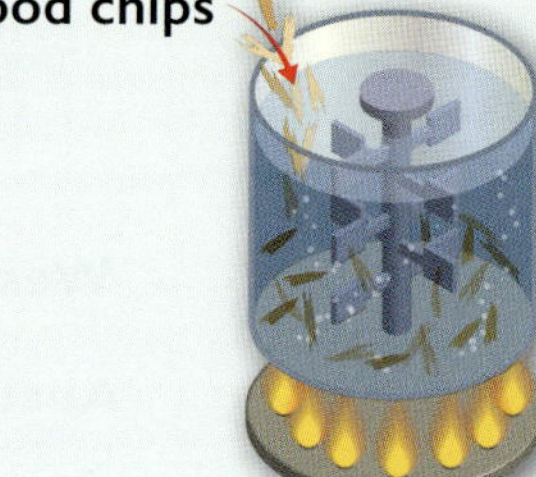

The wood chips are heated in a chemical solution until they turn into a soft pulp.

Iron ore is mined from open-cut mines.

Iron ore is transported to the steel mill.

The iron is melted in a blast furnace. It is then mixed with coal and limestone to make pig iron.

Oil is drilled from under the land and the sea.

Pipelines and large ships transport the oil to a refinery.

The oil is heated to 816 degrees Celsius in huge containers. Ethylene and propylene are produced.

Glass is made from sand.

Cars are man-made. However, they are made from many natural resources.

Tyres are made from rubber.

Plastic is made from oil.

Aluminium is made from bauxite.

discover!

What natural resource is used to make paper?

Where are most of the world's cars made? (page 104 ▶▶)

Which large tower is made of iron? (page 111 ▶▶)

Final product

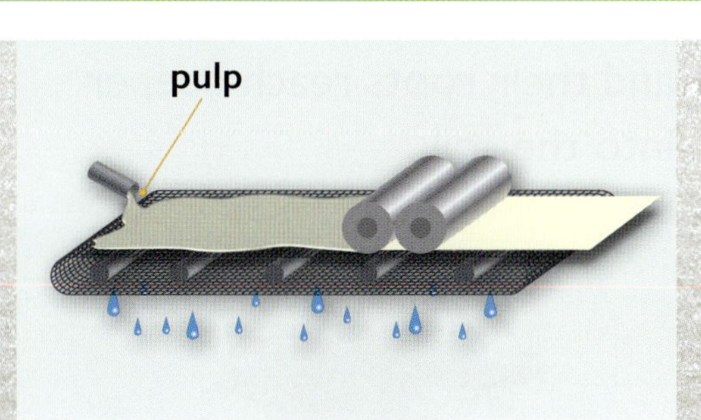

The pulp is poured onto a moving screen and squeezed through rollers to remove the water.

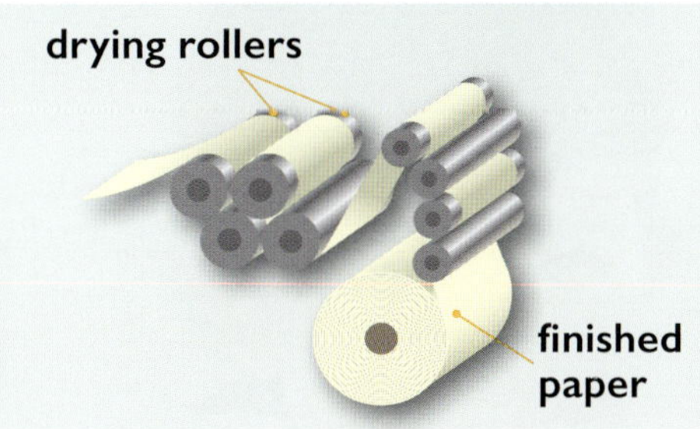

Steam-heated rollers remove the remaining water from the paper. It is then wound onto large reels.

Paper is used to make books, toilet paper and many other products.

Impurities are removed from the pig iron and other materials are added to make the steel strong.

The molten steel is poured into castings to make different shapes. Rollers flatten the steel into sheets.

Steel is a strong material used to make bikes, cars, tools, bridges and skyscrapers.

Chemical reactions turn ethylene and propylene into plastic resin. The resin is cooled and chopped into pellets.

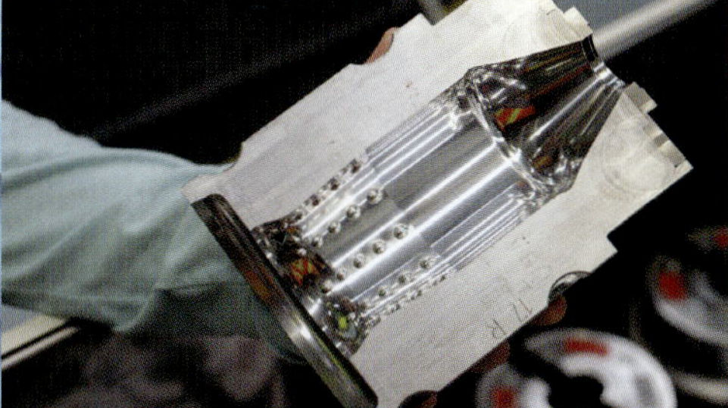

Pellets are melted and poured into molds to make the shapes that we need to make plastic products.

Plastic is a strong, light and waterproof material. It is used to make things like toys and chairs.

Living and non-living things

All animals and plants are living things. Living things feed, breathe, grow, reproduce, move, remove waste and feel change. Cars move with help from the driver. However, a car is not a living thing as it cannot grow or reproduce.

Feeding

All living things take in food to gain energy for their bodies. This allows them to grow and stay healthy.

With a healthy diet, children have plenty of energy to stay active.

Breathing

All living things breathe. Animals breathe in oxygen and breathe out carbon dioxide. Plants breathe in carbon dioxide and breathe out oxygen.

On a cold day, you can see your breath as it turns to water droplets.

Growth

Living things grow and change as they get older. Trees begin as small seedlings. As they get older they grow taller, their trunk becomes hard and wider, and their roots reach deeper into the soil.

The tallest living thing in the world is a 115-metre-tall redwood tree. It is growing on the west coast of the United States of America.

discover!

What do anacondas feed on to gain energy and grow? (page 133 ▶▶)

What is the world's tallest animal? (page 119 ▶▶)

How do plants reproduce?

Reproduction

When living things reproduce, they make new living things. Plants make seeds that grow into new plants. Humans give birth to babies, dogs produce puppies and baby chickens hatch from eggs.

A baby chicken hatches from an egg.

Movement

Living things move material from one part of their body to another. Trees take in water through their roots and move it throughout the plant. Some living things can move from place to place. Animals can run, swim or fly.

The world's fastest land animal is the cheetah. It has been timed at over 90 kilometres per hour. The cheetah lives in southern Africa.

Waste removal

All living things remove waste. Humans remove liquid waste called urine. We also breathe out carbon dioxide to remove it from our bodies.

The white stork urinates on its legs to keep cool in the hot deserts of Africa.

Sensitivity

All living things are sensitive to change. They can feel touch or react to light, heat or sound.

The hairs on the Venus flytrap plant are sensitive. When an insect lands on the hairs, the plant snaps shut, catching the fly.

Extinct animals

Some animals and plants that once lived are now extinct. This means that there are no more of these plants and animals left alive on Earth. We learn about animals and plants that lived a long time ago from their bones and fossils.

▲ A skeleton of a *Diprotodon* dug from Lake Callabonna in South Australia.

Dinosaurs

Dinosaurs became extinct about 65 million years ago. Scientists believe that an asteroid crashed into Earth or a giant volcano erupted. Many dinosaur bones have been discovered. Palaeontologists piece together these bones to build skeletons and to study how the dinosaurs might have lived. Sharp teeth tell the palaeontologists that the dinosaur was a meat eater (a carnivore). Plant eaters (herbivores) had flat teeth for stripping and grinding leaves from the trees.

▲ The *Tyrannosaurus Rex* grew to 12 metres high. Its large, sharp teeth were used for eating meat.

Australian megafauna

After the dinosaurs became extinct, huge mammals began to flourish in Australia. The *Diprotodon* looked like a wombat and was as big as a hippopotamus. These animals are now extinct, but some of them look like animals we find in Australia today. Can you see any animals in the picture that look like Australian animals that are alive today?

discover!

Remains of what extinct animal have been found in northern Siberia?
(page 116 ▶▶)

How did the dodo become extinct?
(page 122 ▶▶)

What Australian megafauna is shown on page 78?

Life cycles

Animals and plants change as they move through their life cycle. Most animals have simple life cycles. Animals are either born alive from their mother or hatched from eggs. The baby then grows slowly to become an adult. Other animals such as frogs and butterflies undergo a big change called metamorphosis.

Plants grow from seeds into small seedlings. When plants mature they produce flowers. The flowers contain seeds to start the cycle again.

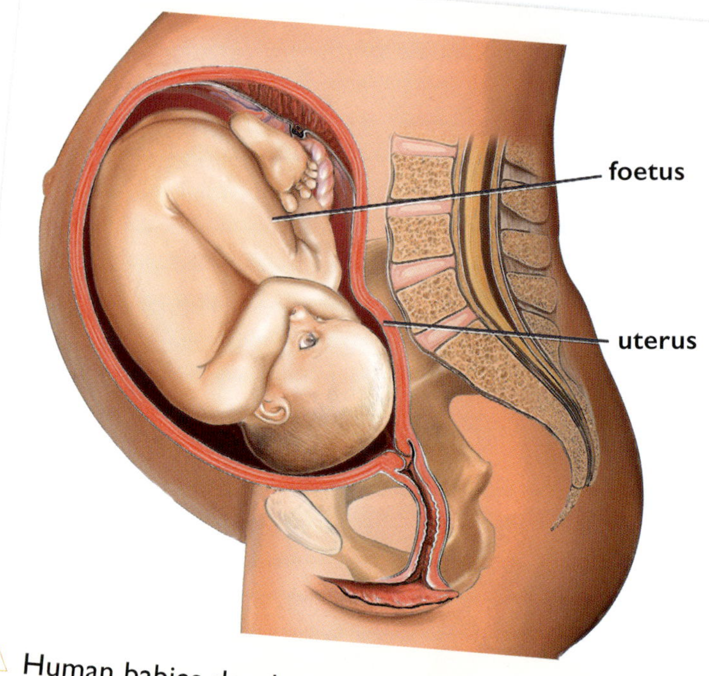

Human babies develop from embryos. An embryo develops into a foetus that grows in the mother's uterus before being born.

Baby saltwater crocodiles hatch from eggs in the Northern Territory.

Hatching from an egg

Reptiles, birds and some mammals hatch their young from eggs. The mother lays eggs in a nest. She sits on the eggs to keep them warm. After some time, the baby hatches from the egg by breaking through the shell.

Metamorphosis

Some animals change completely in their life cycle. They start as one creature and change into another. This is called metamorphosis. A butterfly goes through a metamorphosis during its life cycle from egg to caterpillar to butterfly.

discover!

Which bird lays its eggs in a nest high in eucalypt trees? (page 88 ▶▶)

How tall is a baby giraffe at birth? (page 119 ▶▶)

Which animal is closest to humans? (page 120 ▶▶)

How do emperor penguins raise their chicks? (page 138 ▶▶)

Life cycle of a monarch butterfly

1. An adult butterfly lays its eggs underneath a leaf.

2. A caterpillar (larva) hatches from the egg. It eats leaves and starts growing quickly.

3. When the caterpillar reaches its full size, it attaches itself head down to a twig. In a few hours they change into a pupa.

4. Inside the pupa the caterpillar is transforming. After two weeks in the pupa, a butterfly emerges. It waits for its wings to dry and flies away to begin the life cycle again.

Hottest and coldest

The Sun is Earth's heat source. Some parts of the Earth are hot all year and some are cold. Many places have seasons—summer is the hottest season and winter is the coldest. The temperature is normally hottest when the Sun is high in the sky. This occurs in the middle of the day.

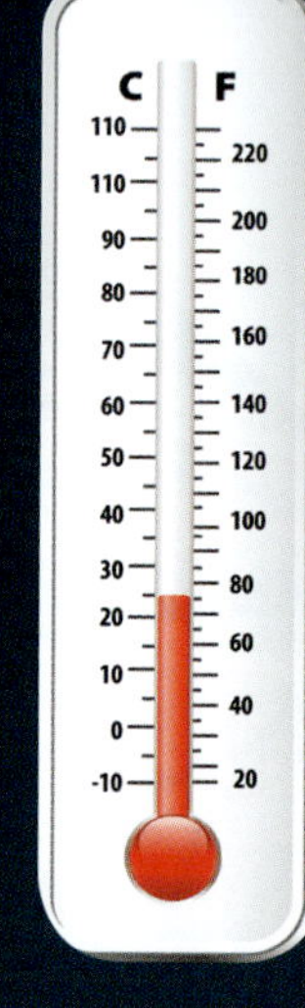

Temperature

Temperature is the measurement of hot and cold. A thermometer is used to measure temperature in degrees Celsius (°C). At 0 degrees Celsius (0 °C), water will freeze. At 100 degrees Celsius (100 °C), water will boil.

World's coldest place

In July 1983, the Russian scientific base Vostok in Antarctica recorded the lowest temperature on Earth. The temperature dropped to –89.6 degrees Celsius (–89.6 °C).

◀ Vostok has a polar climate that is very cold and dry.

World's hottest place

Death Valley, in the United States of America, recorded a temperature of 56.7 degrees Celsius (56.7 °C) in July 1913. It is the hottest temperature ever recorded on Earth.

◀ Death Valley has a desert climate that is very hot and very dry.

Average April temperatures in Sydney ▶

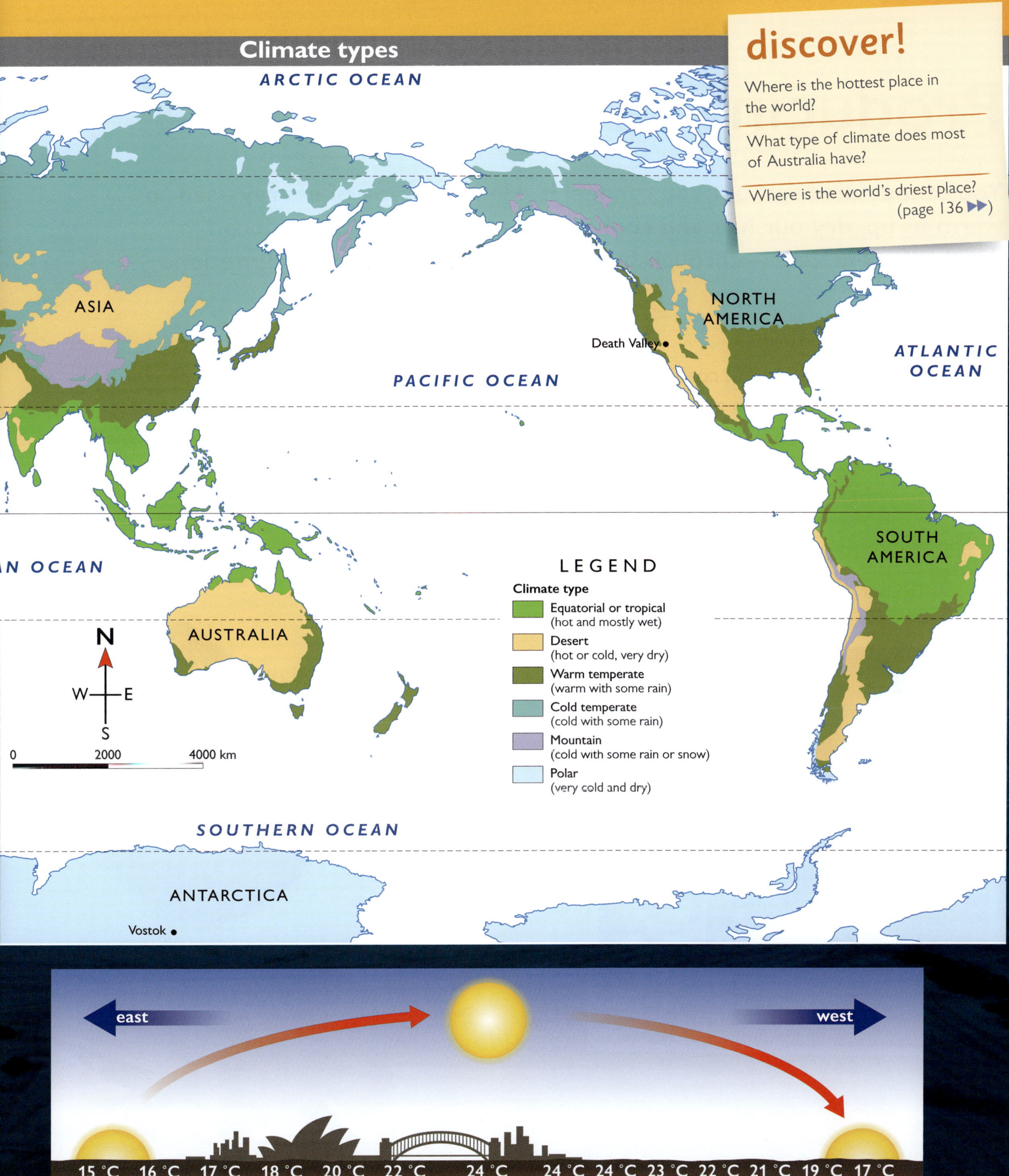

discover!

Where is the hottest place in the world?

What type of climate does most of Australia have?

Where is the world's driest place? (page 136 ▶▶)

Producing heat

You cannot see heat, but you can feel it. Heat is a type of energy used to warm us up, dry our hair and cook our food. Heat can be produced in many different ways—through burning, electricity, friction and motion.

Burning

We burn fuels such as wood, coal or gas to create heat. Fossil fuels such as coal, oil and gas are found under the ground. These fuels have been made over millions of years from the remains of plants and animals.

Wood is burned in a campfire to produce heat. When camping outdoors, we need heat to keep warm. The heat from the fire also helps to cook food and boil water.

Electricity

Electricity produces heat in our home. It is used to produce heat in hairdryers, toasters, heaters and kettles. A metal heating element powers these appliances. The special metal in the heating element resists the flow of electricity and heats up. The heat from the element cooks your toast and boils the water inside a kettle.

The element inside a toaster glows red hot when toasting bread. If toast gets stuck, do not try to remove the toast with your fingers or a knife. Unplug the toaster and call an adult to get the food out.

Friction

When two things rub together, they create heat through friction. If you rub your hands together, the friction creates heat. We can warm our hands on a cold day by rubbing them together.

This type of photograph is called a 'thermogram'. The colours show the heat produced when we rub our hands together. White is the hottest, followed by red, yellow, green, blue and black (the coldest).

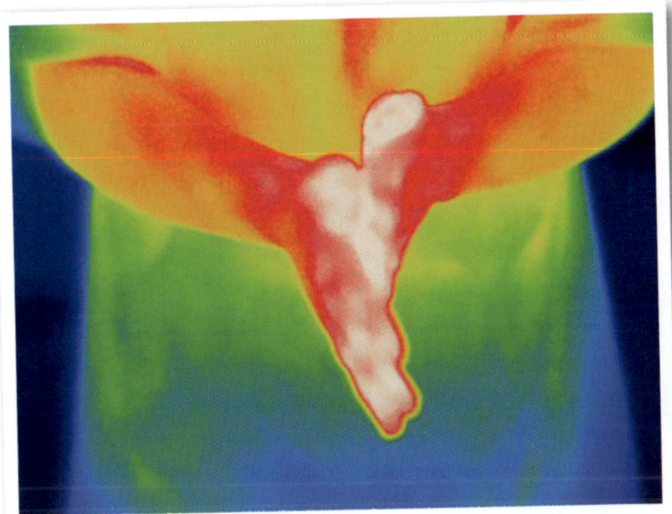

The brakes on a car push against the wheel to create friction to slow it down. The friction also creates heat, as you can see from the smoke on this braking racing car.

discover!

What fuel do we burn to create electricity? (page 80 ▶▶)

How are homes heated in Iceland? (page 112 ▶▶)

What type of tyre is designed to reduce friction? (page 48 ▶▶)

Motion

Humans and other warm-blooded mammals and birds can make heat by moving. When you run, your muscles contract and generate heat energy. You feel hot and you may perspire. If you become too cold, your body automatically creates motion to warm you up. When you shiver, your muscles shake to create heat.

How do we use heat?

Heat moves throughout our environment constantly. It always moves from hotter objects to colder objects. Heat can move from one place to another in three ways: radiation, convection and conduction.

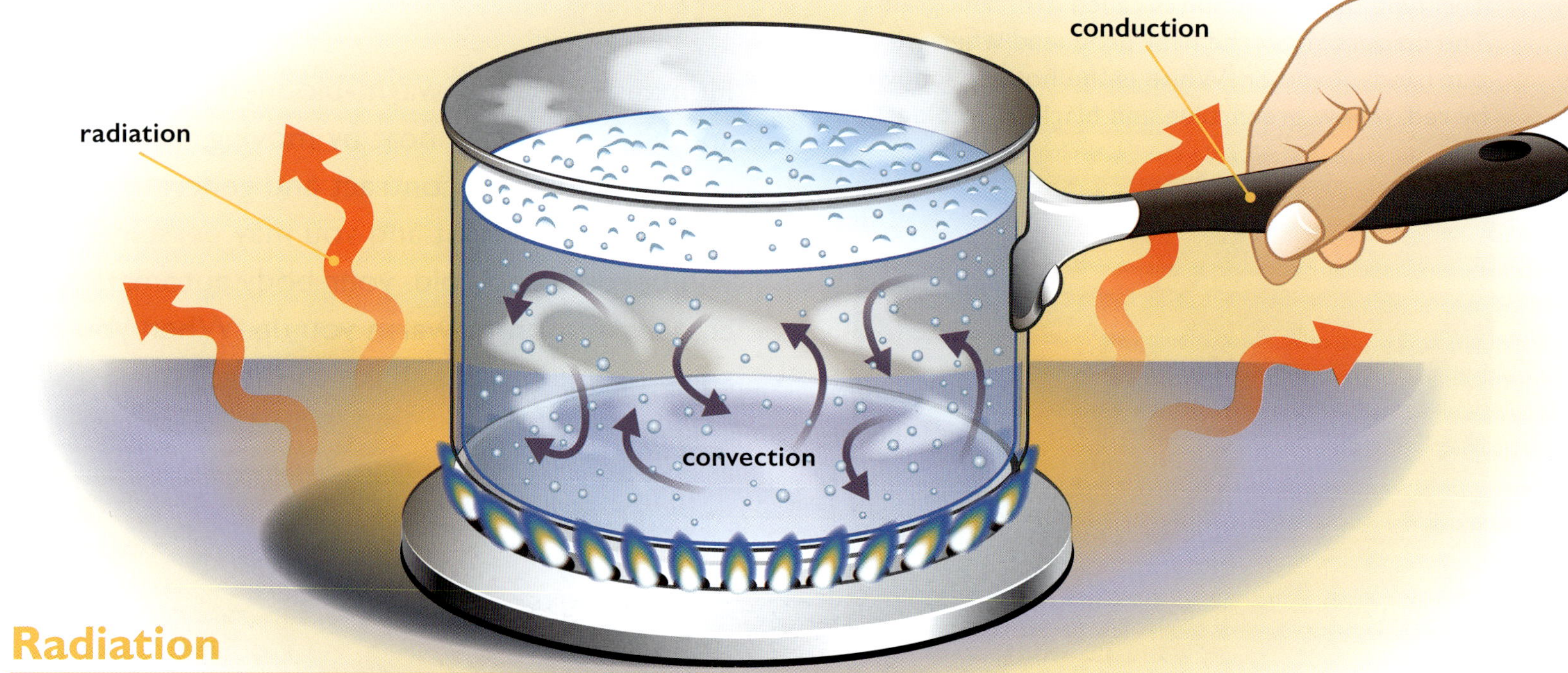

Radiation

Radiation is the movement of invisible waves called rays. The Sun sends out heat in invisible waves. You can feel these rays on your body on a hot day. Fires can also radiate heat to keep us warm.

A fire radiates heat. You can feel the heat if you are sitting close enough.

Convection

Liquids (such as water) or gases (such as air) can carry heat. This process is called convection. In an oven, the air is heated and it moves around to cook or heat food. Heated water from the hot water service moves though the pipes to give you hot water in the kitchen and in the bathroom.

Ovens use convection to cook or heat food.

discover!

How does heat move?

How is conduction helpful for reptiles?

To what temperature is oil heated when making plastic? (page 32 ◀◀)

Conduction

Conduction occurs when two objects are in contact and the heat from one object is transferred to another object. Not all materials conduct heat well. Metal pots and pans are good heat conductors. They heat up food. Wood is not a good conductor. Your hands are protected from heat when you use a wooden spoon to stir a pot of hot soup.

Lizards and other cold-blooded reptiles cannot control their own temperature. They lie on warm rocks to heat their bodies through conduction. The heat from the rock passes to the lizard's body.

Using heat to melt objects

Heat melts solid objects such as chocolate and turns them into liquids. Different solids melt at different temperatures. These are called their melting points.

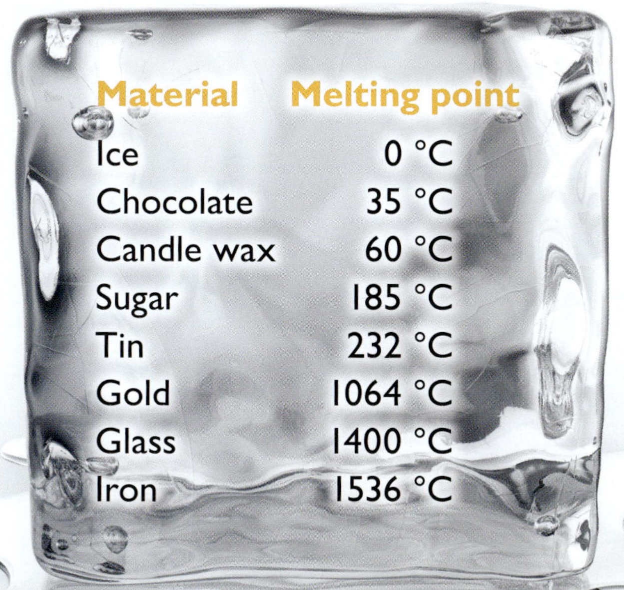

Material	Melting point
Ice	0 °C
Chocolate	35 °C
Candle wax	60 °C
Sugar	185 °C
Tin	232 °C
Gold	1064 °C
Glass	1400 °C
Iron	1536 °C

To make bowls, the glass needs to be melted at a very high temperature. The furnace needs to be heated to 1400 degrees Celsius.

Chocolate melts at just 35 degrees Celsius, so it will melt on a hot day.

Different forces

Forces are pushes and pulls. There are contact and non-contact forces that make things speed up, slow down and change direction.

Contact forces

There are contact forces like pushing a toy car or kicking a ball. The force of a push or a kick moves an object. Friction (page 48) is a contact force between two surfaces that slows things down. Air resistance is a type of friction that slows things down. Walking into a strong wind is an example of air resistance.

Trying to walk into a very strong wind can be difficult. Air resistance is pushing against these walkers.

The force is pushing the tractor.

The tractor moves forward.

Friction between the grass and the tyres slows the tractor down.

Non-contact forces

There are also non-contact forces such as gravity and magnetism (page 50). Non-contact forces push and pull objects without touching them. Gravity is the force that makes things fall to the ground. On Earth, gravity stops everything (including our air) from drifting into space.

discover!

What example of air resistance occurred in the Caribbean in 2012? (page 130 ▶▶)

Why does the log flume ride go so fast? (page 128 ▶▶)

How does the gravitational pull between the Earth and the Moon affect our seas? (page 125 ▶▶)

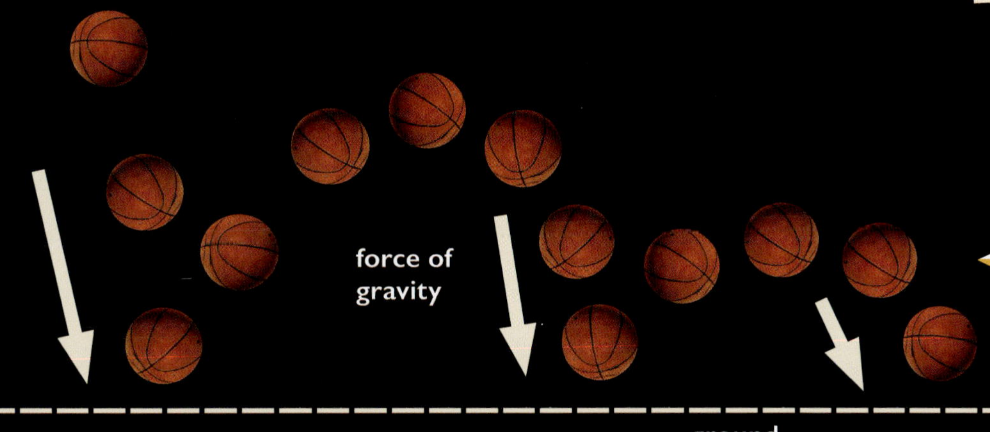

If you drop or throw a basketball, gravity will pull it to the ground.

The force of gravity is much weaker on our Moon. This is because the Moon is much smaller than Earth. Imagine you can jump 30 centimetres high on Earth. On the Moon, you could jump two metres high. You could also throw a ball six times further.

When astronauts walked on the Moon, they hopped across the surface.

Speed and friction

Forces influence the speed at which an object moves. Friction is a contact force between two surfaces that are trying to slide across each other. Friction always slows a moving object down.

This woman has slipped on the ice. Less friction is produced on smooth surfaces like ice.

Useful friction

Friction can be a useful force because it stops our shoes slipping on the floor and helps cars to stop. More friction is produced when the surfaces are rough. Ice is very smooth and causes very little friction. It is easy to slip on ice, but it helps ice-skating because there is little friction.

Tyres are made of rubber. Rubber is waterproof and resists tearing. Tyres are designed to increase or decrease friction with the ground. Less friction increases speed. More friction reduces speed.

This bike tyre is designed to increase friction. It gives the rider greater grip on slippery slopes.

This smooth, thin tyre is for a racing bike. It is designed to reduce friction.

The force of the air pushes against the car.

Air resistance

Air resistance is friction between air and another object. When a car drives along the road, air particles hit the car and slow it down. Cars use smooth, curved (streamlined) shapes to reduce air resistance. Aeroplanes are also streamlined to help them move through the air as easily as possible.

discover!

Does friction speed up or slow down objects?

How do people in Amsterdam overcome friction in winter? (page 111 ▶▶)

How can you use friction to warm your hands? (page 43 ◀◀)

This Australian Olympic cyclist needs to reduce air resistance. She wears a smooth, curved helmet and smooth, tight clothes. The rider bends down to let the air flow over her. The wheels are closed in to help the air flow around them. ▶

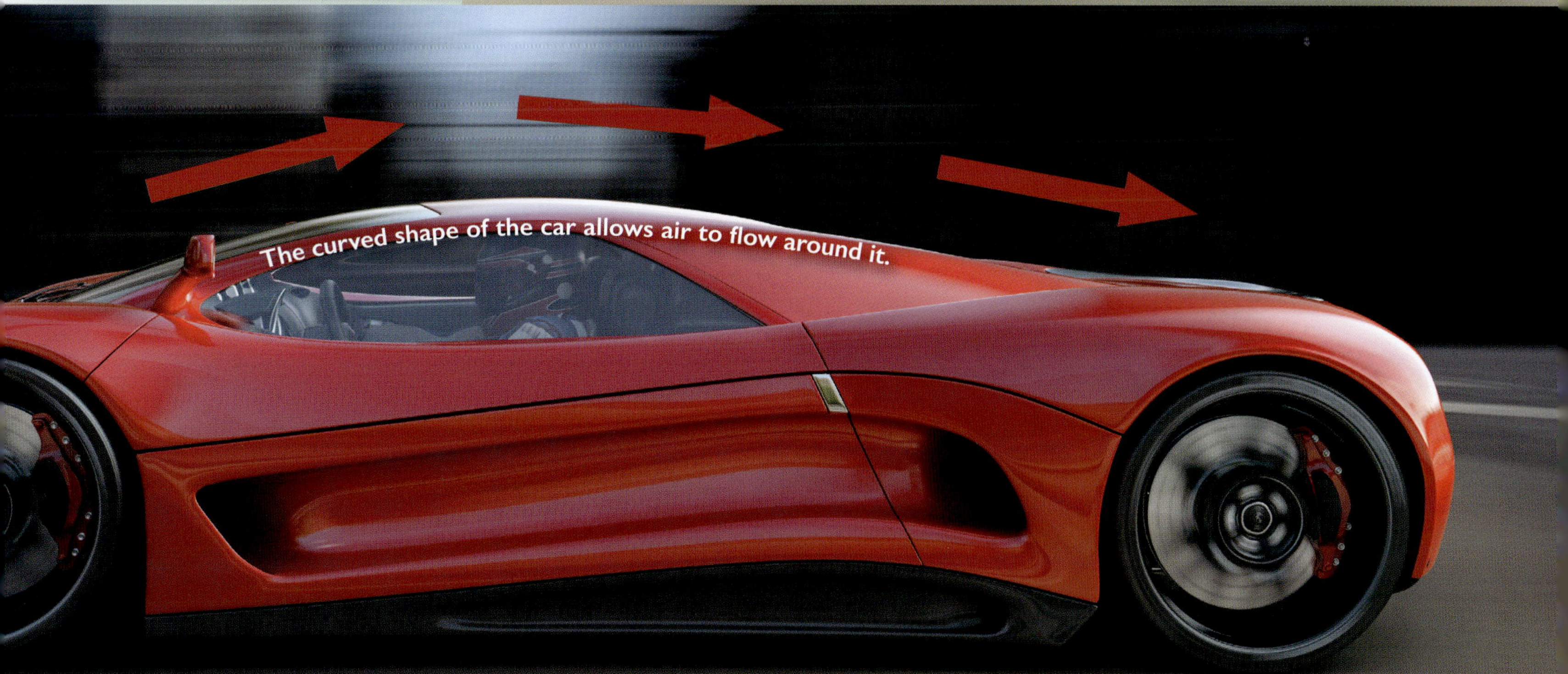

The curved shape of the car allows air to flow around it.

▲ This car is aerodynamically designed to reduce friction.

Magnets

Magnets have an invisible force that can pull or push objects without touching them. Any metal containing iron will be attracted to a magnet.

Magnets on toy trains help connect carriages.

How magnets work

Magnets push and pull one another. The forces are strongest at the ends of the magnets. Magnets have a north pole and a south pole like the Earth. Magnets push away from one another when the same poles are brought close together. We say that same poles repel one another. Magnets pull towards each other when different poles are brought together. We say that different poles attract one another.

Two north poles repel each other.

North and south poles attract each other.

The Earth has a North Magnetic Pole and a South Magnetic Pole. The needle on a compass will always point to the North Magnetic Pole. We use a compass to help find our way. When we know where north is, we can find the other directions (page 6).

discover!

What materials are attracted to magnets?

How are magnets used in junk yards?

Where is the North Magnetic Pole? (page 126 ▶▶)

Using magnets to separate

Metals containing iron are attracted to a magnet. We use magnets to help separate iron from other materials.

The rubbish in your recycling bin contains paper, glass, steel and other materials. At the recycling centre, steel and other metals containing iron are separated using magnets.

This large magnet is in a junk yard. It is being used to separate iron and steel scrap metal from non-magnetic scrap metal.

A horseshoe magnet attracts an iron nail, steel paper clips and screws. It does not attract the plastic building block, the wooden pencil, the coin (zinc and copper) or the brass hose fitting.

Australia's first people

Australia's first people lived in tribes (separate groups), each with their own language, customs and laws. While the majority of Indigenous Australians now live on the east coast of Australia, there are still many Indigenous communities throughout Australia. Aboriginal and Torres Strait Islander people have a strong spiritual connection to the land, and this relationship is important in every aspect of their lives.

Using resources

Trees provided some Aboriginal and Torres Strait Islander people with bark for shelter, canoes and shields. They also used wood from trees for spears, digging sticks, throwing sticks, boomerangs and bowls.

Traditionally, the women were the main food gatherers. They searched for food such as seeds, vegetables, fruit and witchetty grubs. These skills have been passed on to modern Aboriginal people.

Honey pot ant

This canoe tree in the Hattah–Kulkyne Nation on the Murray River shows how early Aboriginal people used natural resources.

A digging stick is being used by this Aboriginal woman. She is looking for a nest of honey pot ants.

Aboriginal stories and legends have been passed down through dance and art.

Hunting for food

Aboriginal and Torres Strait Islander people were skilled trackers and hunters. They fished the rivers and hunted kangaroos, wallabies, birds and lizards.

Aboriginal people used boomerangs to hunt animals.

discover!

What animals do Xingu Indians hunt? (page 132 ▶▶)

'The land is my mother, my mother is the land.' What does this mean?

Who was William Buckley and what did he learn from the Aboriginal people? (page 86 ▶▶)

Connection to the land

Aboriginal people live in harmony with the land.

'We don't own the land, the land owns us. The land is my mother, my mother is the land. Land is the starting point to where it all began. It's like picking up a piece of dirt and saying this is where I started and this is where I'll go. The land is our food, our culture, our spirit and identity.'

Knight S., 1996,
Our Land, Our Life

Aboriginal languages

Aboriginal people pass on their cultural identity, traditions and laws through stories, songs, art and language. At the time of the European settlement in 1788, there were over 250 Aboriginal languages spoken across Australia. The number of languages spoken now is less than half—approximately 120. Many indigenous communities are working hard to protect their unique languages by recording them and teaching them to the next generation.

Indigenous sustainability practices

Aboriginal and Torres Strait Islander people have a deep knowledge of the land. This special relationship is important in every aspect of their lives. They use both traditional and modern sustainability practices to manage the natural resources and environments.

Fire

Aboriginal people use fire to look after and heal their country. At the right time of the year and day, slow-burning fires are lit to manage the growth of grasses and scrub and to rejuvenate the land. These cool burns are often managed by Aboriginal rangers and have a lower impact on wildlife and flora than a raging bushfire.

Art

Ancient Aboriginal rock art tells stories and describes cultural practices and the environment. Pieces of rock art discovered in Arnhem Land in the Northern Territory date back to 28000 years ago. This is some of the oldest rock art in the world. The conservation of this art is an important part of sustaining Aboriginal history.

▲ This rock art, depicting a fish, is found in Ubirr, Northern Territory.

◄ A ranger participates in cool burning in the Wardekken Indigenous Protected Area, in Arnhem Land, Northern Territory.

Waterways

Many of Australia's waterways are special to Aboriginal people and need to be protected. There are many significant sites along, and in, river beds. Making sure there is no overuse of these waterways will help to protect these environments.

discover!

How is fire used as a sustainability practice?

What other kind of Indigenous rock art is found in Australia? (page 74 ▶▶)

What sustainability practices would have helped this extinct Mauritian species? (page 122 ▶▶)

▲ The Coorong National Park in South Australia is a protected area and very special to Indigenous Australians.

Threatened species

Aboriginal leaders work with state and federal governments to protect threatened species, such as dugongs. This work includes protecting habitats, tagging and tracking wildlife, and setting up parks to protect threatened species.

Rangers patrol the waters off the Gulf of Carpentaria, removing rubbish that has washed into the sea. ▶

Exploration

Humans have always explored the world. Explorers recorded their expeditions as maps. The maps continued to change as new information about the world was discovered. Early sea explorers believed that the world was flat. They thought they would fall off the edge of the world if they sailed too close to the horizon, so they tended to stay close to coastlines.

James Cook

Captain James Cook explored and mapped the east coast of Australia in his ship the *Endeavour*. In 1770, Cook named Botany Bay (now Sydney). Cook's crew fired warning shots at two Aboriginal people who responded by throwing spears. Even though the Aboriginal people inhabited Australia, Cook claimed the whole eastern coast for the King of England, naming it New South Wales.

Leif Eriksson

Leif Eriksson was a Viking. Eriksson was also the first European to reach North America. In 1001, he established a small settlement in modern-day Newfoundland. His discovery of this new world remained a secret and most of the credit for North America's discovery went to Christopher Columbus, nearly 500 years later.

Exploration and discovery dates

500 BCE	120 CE	982	1001	1271–95	1405–33	1487–88	1492–93	1519–22
The Silk Road, the trade route between Europe and Asia, is established.	Ptolemy makes the first flat map of the world.	Eric the Red discovers Greenland.	Leif Eriksson	Marco Polo goes to China.	Zheng He sails from China to the Pacific islands, the Middle East and Africa.	Bartholomew Diaz rounds the Cape of Good Hope.	Christopher Columbus sails to America.	Ferdinand Magellan sails around the world.

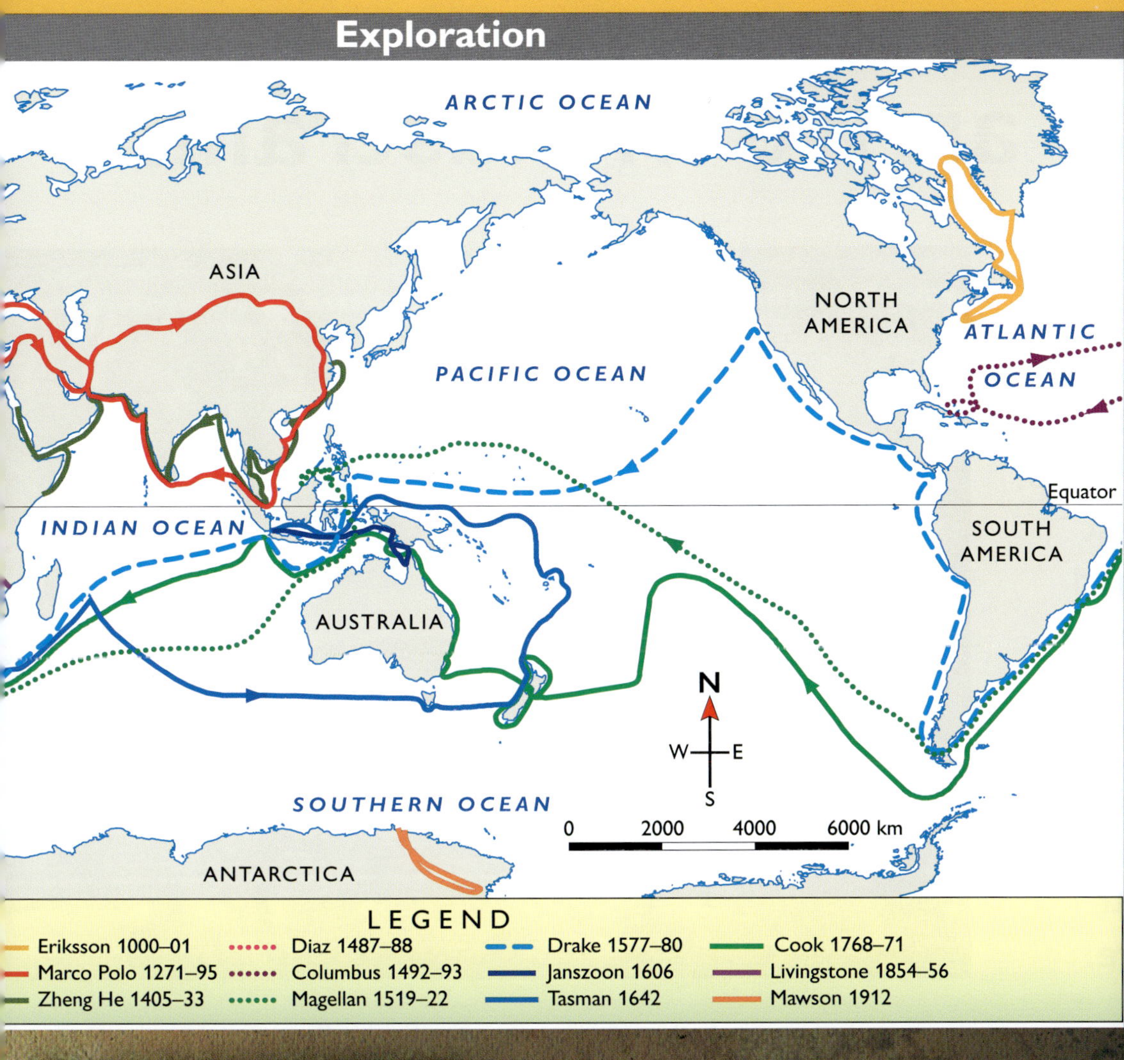

discover!

What happened when Cook first met Aboriginal people in Australia?

Why did Stanley say 'Dr Livingstone, I presume?'?

What happened when Abel Tasman met Māoris in New Zealand?

(page 98 ▶▶)

Sir Douglas Mawson

In 1912, Australian Sir Douglas Mawson, Xavier Mertz and Belgrave Ninnis explored an unknown region of Antarctica. After travelling over 500 kilometres across the ice, Ninnis and Mertz both died. Mawson then battled frostbite and starvation to reach the expedition base.

Dr David Livingstone

In 1856, Scotsman Dr David Livingstone became the first European to travel across Africa. After being reported missing, Livingstone was found by Henry Morton Stanley. 'Dr Livingstone, I presume?', said Stanley when he saw Livingstone on the shores of Lake Tanganyika in 1871.

1577–80 Francis Drake ails around the world.

1606 Willem Janszoon discovers the northern coast of Australia.

1642 Abel Tasman reaches Van Diemen's Land (Tasmania) and New Zealand.

1770 James Cook

1856 David Livingstone

1909 Robert Peary discovers the North Pole.

1911 Roald Amundsen is the first person to reach the South Pole.

1912 Douglas Mawson

1961 Yuri Gagarin is the first man in space.

1969 Neil Armstrong is the first man on the Moon.

The British arrive in Australia

On 18 January 1788, the First Fleet arrived at Botany Bay. They were to establish a colony to take convicts from Britain's prisons. Captain Arthur Phillip, the commander of the First Fleet, found Botany Bay had no fresh water so the fleet relocated to Port Jackson. The 751 convicts and their children, and 252 marines and their families came ashore to begin Australia's first white settlement at Sydney Cove.

Captain Arthur Phillip raised the British flag at Sydney Cove on 26 January 1788. This day is celebrated as Australia Day. Many Aboriginal people call it Invasion Day.

First contact

When Aboriginal people saw the white skinned people of the First Fleet they thought they were the spirits of dead ancestors. Aboriginal people and new settlers had very different ideas. They did not understand the new settlers' idea of ownership and they believed that sheep and cattle were for everyone to share.

When the Aboriginal people killed the livestock, the new settlers fired their guns at them. The Aboriginal people then threw spears at the settlers.

Groups of Aboriginal people in the Sydney area

New South Wales
Sydney
Broken Bay
Nepean River
Sydney Harbour
Botany Bay
N
W
E
S
0 10 20 km

LEGEND

Aboriginal group

1 Kurrajong
2 Cattai
3 Boorooberongal
4 Bidjigal
5 Toogagal
6 Gomerrigal
7 Cannemegal
8 Mulgoa
9 Bool-bain-ora
10 Cabrogal
11 Muringong
12 Carigal
13 Cannalgal
14 Borogegal
15 Kayimai
16 Terramerragal
17 Cammeraigal
18 Gorualgal
19 Birrabirragal
20 Cadigal
21 Burramattagal
22 Wallumattagal
23 Wangal
24 Muru-ora-dial
25 Kameygal
26 Bediagal
27 Gweagal
28 Tagary
29 Norongerragal

In 1788, there were thought to be more than 350 000 Aboriginal people in Australia. There were 29 different groups of Aboriginal people in the Sydney area.

discover!

Why did the Aboriginal people think the First Fleet were spirits?

Why was Governor Phillip speared?

What building is located at Bennelong Point? (page 73 ▶▶)

Which convict escaped and lived with the Aboriginal people for 32 years? (page 86 ▶▶)

Bennelong

Bennelong was one of the first Aboriginal people to live with the white settlers. He came from the Wangal people. Bennelong was captured in November 1789. He learned English and taught Phillip some Aboriginal customs. Phillip built Bennelong a hut on what is now known as Bennelong Point, the site of the Sydney Opera House.

Phillip is speared

Governor Arthur Phillip was invited to join a feast by his Aboriginal friend, Bennelong. Phillip went to shake the hand of a tribesman. The tribesman thought Phillip was trying to grab him. He speared Phillip in the shoulder.

Governor Phillip (in the centre of the beach) escapes to the boats with the spear in his shoulder.

Living cultures

In Australia, we embrace our cultural diversity and unique history. The Indigenous cultures of Australia are the oldest living cultures in the world. They go back at least 65,000 years. Learning about these diverse cultures helps us to better understand our history as Australians.

Aboriginal and Torres Strait Islander flags

The Aboriginal flag was designed by Aboriginal elder Harold Thomas in 1971. The black represents the Aboriginal people, the red represents the earth, and the yellow represents the Sun.

The Torres Strait Islander flag was designed by Bernard Namok in 1992. The green represents the land, the black represents the people, the blue represents the sea, and the white represents peace.

NAIDOC Week

NAIDOC Week is held in July and celebrates Aboriginal and Torres Strait Islander history and culture. It is a time to recognise the contributions that Indigenous Australians make to Australian society. There are many NAIDOC week celebrations around Australia including dancing, visual arts events, community programs and awards.

People participate in a NAIDOC week march.

discover!

Why is NAIDOC Week important?

How did the Māori people react when the first white explorer reached New Zealand?

(page 98 ▶▶)

Why is it important to show respect for the traditional owners of the land?

Anniversary of the National Apology

On 13 February 2008, the prime minister Kevin Rudd delivered his apology speech to the Stolen Generations. This speech acknowledged the experiences of Aboriginal and Torres Strait Islander children who were removed from their families.

Welcome to Country

At many official events and ceremonies, a Welcome to Country is performed. This is a ceremony conducted by local Indigenous Elders to welcome visitors and to show respect to current and former custodians of the land. The ceremony can be a speech, song, dance or smoking ceremony. Only Elders who represent the Traditional Owners of the land can perform a Welcome to Country.

An Indigenous Elder performs a smoking ceremony at the opening of a new school campus.

Communities around the world

People around the world are different in many ways. They have different skin colours. They speak different languages. They have different religions. However, people are also the same in many ways. They live in communities. They have families. They go to work, school and special events in their communities.

People at work

People undertake many different jobs and in some communities children work to help their families.

▼ This mother and son are harvesting wheat by hand in Turkey.

▲ This combine harvester is harvesting wheat in Australia.

People at play

People around the world play a variety of different games. Some of these games depend on the weather.

▼ These children are playing cricket in India.

▲ These children are playing ice hockey in Canada.

People and culture

The culture and beliefs of people can be seen in their dress, ceremonies and the foods they eat.

This mother from Papua New Guinea helps her son get ready for a ceremony. ►

People living together

People live in different settlements—from small villages to towns, regional centres and large cities.

The small village of Lower Slaughter in England has a population of 200 people. ►

These Native American children are in ceremonial dress.

discover!

Name some foods that have been brought to Australia by people from different countries. (page 66 ▶▶)

What is the name of the traditional Māori war dance? (page 98 ▶▶)

What do the Uros people use to build their homes? (page 134 ▶▶)

These boys from Alaska (USA) use a snowmobile for their transport.

People on the move

People move from place to place. They need transport to get to work or school. We use aeroplanes to travel long distances.

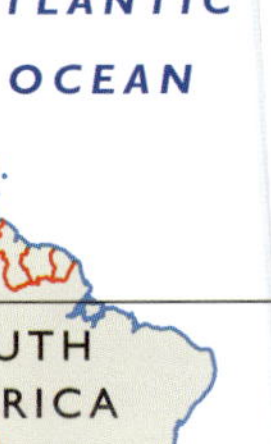

Japan's bullet trains travel at up to 300 kilometres per hour.

People at home

Houses are made of different materials to give shelter from hot, cold and wet weather. Houses provide homes for people.

Sao Paulo is a large city in Brazil where more than 21 million people live.

Most people in Singapore live in tall apartment blocks.

This home in Africa is made from wood, clay and straw to keep the family cool.

Democracy

Democracy means 'rule by the people'. In a democracy, most people have a say or a vote. When we vote for something and are included in making decisions, it helps us to feel valued and take responsibility for making something the best it can be. Usually, in a vote, the side or idea with the most votes wins.

Australian citizens aged 18 years and over must vote on election days. They vote for someone who best represents their opinions in the House of Representatives and the Senate.

discover!

What does the word 'democracy' mean?

What are the three forms of government in Australia?

Where did the idea of democracy come from? (page 110 ▶▶)

A democratic system of government

Australia has a democratic system of government. This means that representatives of the government are elected by the people in a public vote.

◀ Government representatives attend a debate.

Federal, state and local governments

Australia has three forms of government: federal, state and local. The federal government looks after issues on a national scale like immigration and trade. State governments look after issues within a state, such as health and education. Your local government looks after the area you live in and is in charge of services such as libraries, parks, cultural events and recreation.

▼ There are also laws to keep us safe on the roads. Here, a car stops at a stop sign to let a train go past.

Laws and rules

One of the main roles of federal and state governments is to make laws. Laws make sure that our society operates safely and effectively. There are usually consequences if laws are broken. Rules, which you might have at home or school, help us understand the correct behaviour in certain situations and places.

Multicultural Australia

One in four people in Australia were born in a different country. People from many different countries live in Australia. Through their cultures, they have introduced different languages, religions, foods and celebrations to our country.

Different languages

The people of the world speak over 6000 languages and dialects. In Australia, English is the main language. Other common languages are Mandarin, Italian, Punjabi, Arabic, Vietnamese and Greek.

Different foods

As people from different countries came to Australia, they brought delicious foods from their home countries for us to enjoy.

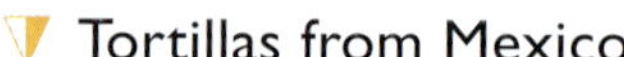

Tortillas from Mexico

Noodles from China and other Asian countries

Pizza from Italy

discover!

Why is dance so important to the different Aboriginal groups in Australia? (page 52 ◀◀)

Which religion does the world's tallest statue honour? (page 100 ▶▶)

Why are so many people bathing in the Ganges River in India? (page 106 ▶▶)

Different celebrations

Our multicultural population celebrates a range of important events. These are times of prayer, gift giving, feasts and tradition.

Ramadan

Ramadan is the Islamic month of fasting. During Ramadan, Muslims don't eat or drink in daylight hours. They learn about patience and to think about others. Muslims also read their bible, the Qur'an, during Ramadan.

These women are serving dinner at their mosque (church) after finishing Ramadan.

Spectacular Chinese dragons at Chinese New Year

Chinese New Year

Chinese New Year starts on the first day of the Chinese calendar, usually in late January or early February. The celebrations last for 15 days. For families it is a time of feasting and celebrating. There are colourful processions with dancers, music and Chinese dragons.

Red envelopes containing money are given at Chinese New Year. The red colour means good luck and is said to frighten off evil spirits.

Remembering the ANZACs

ANZAC Day is commemorated each year on 25 April. It marks the anniversary of the first war fought by Australian and New Zealand forces in 1915. ANZAC stands for Australian and New Zealand Army Corps. On ANZAC Day, we remember all the men and women who have fought in wars to keep Australia safe.

Symbols of ANZAC Day

The Rising Sun

People buy pins to raise money for returned soldiers on ANZAC Day. The badge features the Rising Sun logo that is worn by all Australian soldiers.

The Australian flag

The Australian flag is flown at half-mast as a sign of respect for those soldiers who fought for Australia.

At the Dawn Service, people stand in silence to remember our soldiers. A bugler then plays the Last Post, a bugle call used to commemorate those who have lost their lives at war.

Children wave Australian flags as soldiers march in an ANZAC parade.

The Rising Sun badge is shown on this hat from the First World War (1914–18).

The ANZAC Day pin features the Rising Sun.

Red poppy

discover!

What are two symbols of ANZAC Day?

Where is the national ceremony for ANZAC Day held every year? (page 84 ▶▶)

Who did the Australian soldiers fight against at Gallipoli? (page 114 ▶▶)

Red poppies

Red poppies are the flower of remembrance worn on ANZAC Day. Red poppies are a reminder of the flowers that grew on the battlefields during the First World War.

Simpson and his donkey

In the First World War, Private Simpson picked up wounded soldiers and carried them on his donkey. He took the wounded back to the hospital with guns firing around him.

This painting shows Private Simpson and his donkey at Anzac Cove.

DISCOVERING OUR COUNTRY

Australia

flag
Australian

coat of arms
Australian Government

Australia is the only populated country that is also a continent. It is divided into six states and two territories. Australia has a population of just 25 million people.

Australia's largest bird

The emu is the largest bird in Australia. It grows to two metres in height. The emu is on our coat of arms, along with a kangaroo. Emu eggs are among the largest eggs in the world. They can measure up to 15 centimetres in length.

Eggs (page 38)

Australia facts

Population: 24 114 200

Largest state (area): Western Australia 2 529 875 sq km

Largest state (population): New South Wales 7 726 000

Largest city (population): Sydney 4 921 000

ARAFURA SEA
TIMOR SEA
INDIAN OCEAN
Great Australian Bight
AUSTRA
Western Australia
Northern Territory
Tropic of Capricorn
Darwin
Palmerston
Jabir
Katherine
Wadeye
Wyndham
Kununurra
Daly Waters
Newcastle Waters
Derby
Broome
Halls Creek
Tennant Cre
Tanami
Barrow Cree
Port Hedland
Karratha
Exmouth
Newman
Alice Spring
Kaltukatjara
Carnarvon
Ernabella
Meekatharra
Kalbarri
Laverton
Geraldton
Menzies
Maralin
Cook
Kalgoorlie
Eucla
Cedu
Merredin
Wanneroo
Northam
Norseman
Perth
Rockingham
Mandurah
Esperance
Bunbury
Katanning
Busselton
Augusta
Albany

N
W E
S

0 250 500 750 km
1 centimetre on the map measures 160 kilometres on the ground.

LEGEND

——	Country border
- - -	State/territory border
Tasmania	State/territory name
□ **Adelaide**	Over 1 000 000 people
○ **Geelong**	50 000 to 1 000 000 people
○ Albany	5000 to 50 000 people
○ Bourke	Under 5000 people
■ ● ● ●	Country capital city
□ ○ ○ ○	State/territory capital city

PAPUA NEW GUINEA

Nhulunbuy
Weipa
PACIFIC OCEAN
Gulf of Carpentaria
Numbulwar
Borroloola
Cooktown
Cairns
Innisfail
CORAL SEA
Burketown
Normanton
ern
ory
L I A
Camooweal
Townsville
Mount Isa
Charters Towers
Bowen
Hughenden
Mackay
Queensland
Winton
Longreach
Yeppoon
Emerald
Rockhampton
Gladstone
Finke
Windorah
Birdsville
Bundaberg
Hervey Bay
Charleville
Roma
Gympie
Oodnadatta
Sunshine Coast
Brisbane
Toowoomba
Ipswich
Gold Coast
Cunnamulla
Tweed Heads
Warwick
th
Coober Pedy
Kati Thanda-Lake Eyre
alia
Moree
Ballina
Lismore
Andamooka
Leigh Creek
Bourke
Inverell
Grafton
River
Narrabri
Coffs Harbour
New South Wales
Armidale
Tamworth
Broken Hill
Port Macquarie
Port Augusta
Dubbo
Muswellbrook
Taree
Whyalla
Port Pirie
Darling
Maitland
Newcastle
Orange
Wyong
Cummins
Berri
Mildura
Bathurst
Gosford
Sydney
Gawler
Hay
Griffith
Adelaide
Port Lincoln
Wagga Wagga
Goulburn
Wollongong
Murray Bridge
Victor Harbor
Murray River
Albury
Canberra
ACT
Batemans Bay
PACIFIC OCEAN
Bordertown
Horsham
Shepparton
Wangaratta
Naracoorte
Bendigo
Victoria
Mount Gambier
Ballarat
Melbourne
Bairnsdale
Geelong
Portland
Moe
Traralgon
Warrnambool
Bass Strait
TASMAN SEA
Burnie
Devonport
Launceston
Queenstown
Tasmania
Hobart

Australia's Gold Coast

The Gold Coast attracts millions of tourists every year. It is famous for its beaches. There are also many attractions on the Gold Coast such as Sea World (pictured), Movie World and Dreamworld.

Australia's largest city

Sydney is Australia's largest city. It is located on Sydney Harbour. The famous Sydney Opera House is built on Bennelong Point.

Who was Bennelong? (page 59)

Western Australia 

coat of arms
Western Australian Government

floral emblem
red and green kangaroo paw

animal emblem
numbat

bird emblem
black swan

World's largest collection of rock carvings

It is estimated that there are between 500 000 and 1 million Aboriginal rock carvings (called petroglyphs) scattered across the Burrup Peninsula and islands of the Dampier Archipelago. Created by Aboriginal people up to 30 000 years ago, these carvings show both land and sea animals—including whales, kangaroos, emus, turtles, fish and birds—as well as Aboriginal ceremonies.

Indigenous sustainability practices (page 54)
Australian megafauna (page 37)

This rock carving of the extinct Tasmanian tiger is very important. It is the only record of its existence in this part of Australia.

The Burrup rock carvings include possibly the first representation of human faces in history.

Western Australia facts

Capital city: Perth, B2

Highest point: Mt Meharry, B4 (1253 m)

Longest river: Gascoyne, A4 (834 km)

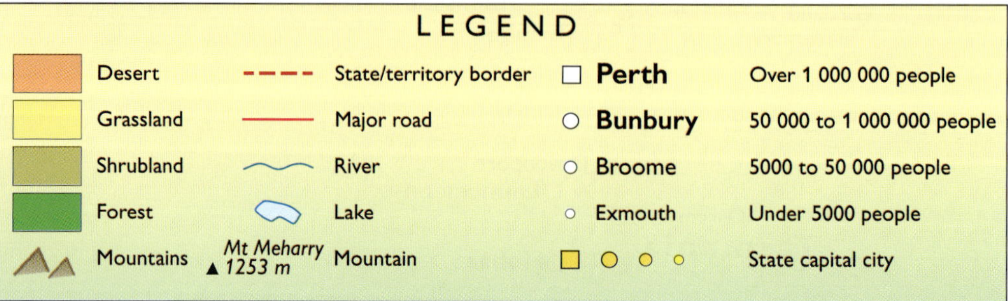

TIMOR SEA
INDIAN
OCEAN
Cape Londonderry
Kalumburu
Joseph Bonaparte Gulf
Wyndham
Kununurra
Drysdale River
Durack River
Ord River
Cape Leveque
Lombadina
King Sound
Kimberley
Lake Argyle
Derby
KING LEOPOLD RANGE
Mt Wells 983 m
BUNGLE BUNGLE RANGE
Broome
Fitzroy River
Fitzroy Crossing
Halls Creek
North West Shelf
Lake Gregory
Port Hedland
Great Sandy Desert
Barrow Island
Karratha
Marble Bar
North West Cape
Exmouth
Onslow
Fortescue River
HAMERSLEY RANGE
Percival Lakes
Mt Bruce 1235 m
Tom Price
Pilbara
Mt Meharry 1253 m
Mt Newman 1053 m
Newman
Lake Mackay
Kiwirrkurra
Tropic of Capricorn
KENNEDY RANGE
Ashburton River
Lake Disappointment
Lake MacLeod
Western
Australia
Gascoyne River
CARNARVON RANGE
Gibson Desert
Carnarvon
Shark Bay
Steep Point
Mt Deering 1219m
Murchison River
Lake Carnegie
Mt Aloysius 1085 m
Meekatharra
Wiluna
Lake Wells
Cue
Lake Austin
Kalbarri
Mount Magnet
Sandstone
Great Victoria Desert
Laverton
Geraldton
Lake Carey
Lake Barlee
Menzies
Kalgoorlie
NULLARBOR PLAIN
Rawlinna
Wanneroo
Perth
Rockingham
Mandurah
Northam
York
Merredin
Lake Cowan
Norseman
Eucla
Hyden
Great Australian Bight
Bunbury
Busselton
Katanning
Ravensthorpe
Esperance
Hopetoun
Margaret River
Augusta
Cape Leeuwin
Flinders Bay
Manjimup
Bluff Knoll 1096 m
Albany
King George Sound
West Cape Howe
Northern Territory
South Australia
N
W
E
S
0 100 200 300 km
1 centimetre on the map measures 100 kilometres on the ground.
110°E
115°E
120°E
125°E
130°E
15°S
20°S
25°S
30°S
35°S
A
B
C
D
E
1
2
3
4
5
6

Northern Territory

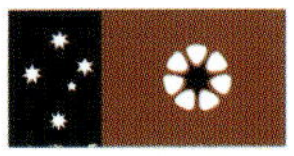

coat of arms
Northern Territory Government

floral emblem
Sturt's desert rose

animal emblem
red kangaroo

bird emblem
wedge-tailed eagle

Jim Jim Falls in Kakadu National Park during the wet season

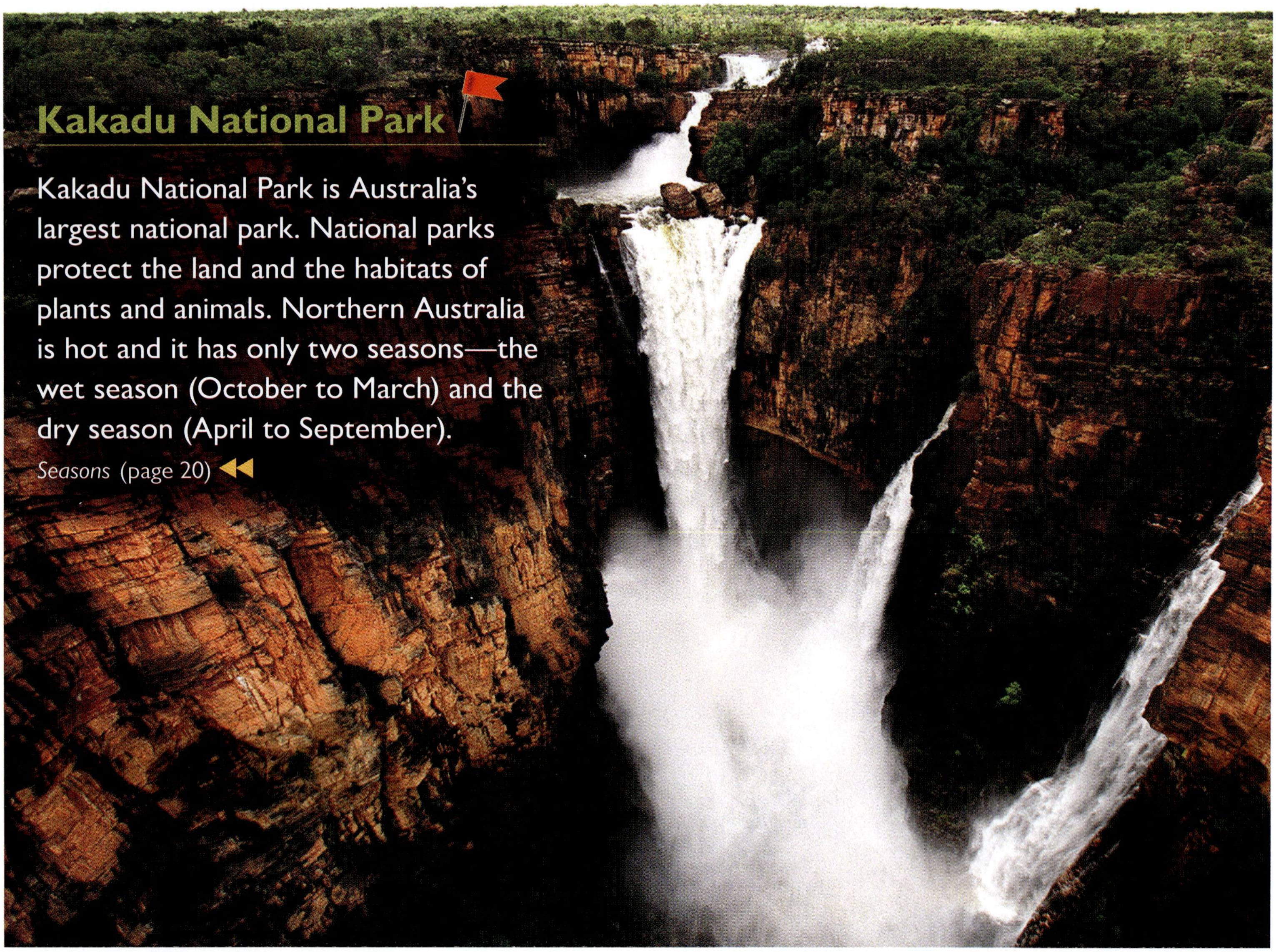

Kakadu National Park

Kakadu National Park is Australia's largest national park. National parks protect the land and the habitats of plants and animals. Northern Australia is hot and it has only two seasons—the wet season (October to March) and the dry season (April to September).

Seasons (page 20)

Northern Territory facts

Capital city: Darwin, C7

Highest point: Mt Zeil, D2 (1531 m)

Longest river: Victoria, C5 (510 km)

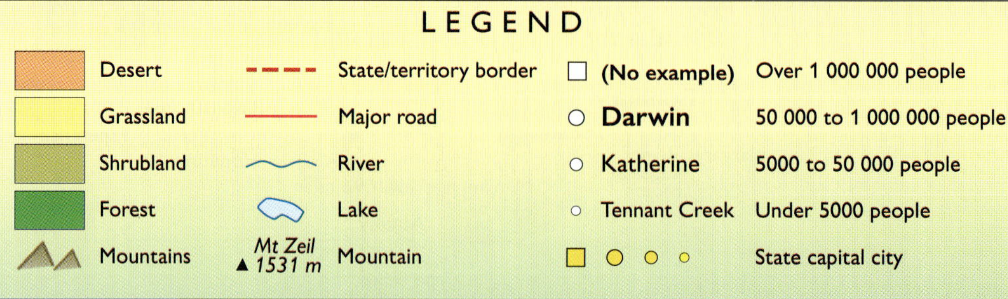

ARAFURA SEA
TIMOR SEA
Cape Van Diemen
Tiwi Islands
Milikapiti
Melville Island
Bathurst Island
Cobourg Peninsula
Van Diemen Gulf
Warruwi
Maningrida
Wessel Islands
Cape Wessel
The English Company's Islands
Nhulunbuy
Cape Arnhem
Gove Peninsula
Gapuwiyak
Darwin
Palmerston
Oenpelli
Jabiru
Arnhem Land
Mt Gilruth 558 m
Adelaide River
Pine Creek
Daly River
Katherine River
Bulman
Groote Eylandt
Angurugu
Numbulwar
Joseph Bonaparte Gulf
Wadeye
Katherine
Limmen Bight
Gulf of Carpentaria
Turtle Point
Roper River
Mataranka
Timber Creek
Larrimah
Sir Edward Pellew Group
Victoria River
Borroloola
Daly Waters
Newcastle Waters
Kalkarindji
Elliott
BARKLY TABLELAND
Lake Woods
Nicholson River
Northern Territory
Tanami Desert
Tennant Creek
Tanami
Devils Marbles 433 m
Western Australia
Queensland
Barrow Creek
Mt Gardiner 999 m
Ti-Tree
Lake Mackay
Central Mount Wedge 1067 m
Mt Brassey 1216 m
Mt Liebig 1267 m
MACDONNELL RANGES
Mt Zeil 1531 m
Tropic of Capricorn
Lake MacDonald
Alice Springs
Mt Conway 1133 m
Todd River
Hay River
Finke River
Lake Amadeus
Kaltukatjara
Kata Tjuta 1069 m
Uluru 868 m
PETERMANN RANGES
Finke
Simpson Desert
South Australia
N
W
E
S
0 50 100 150 km
1 centimetre on the map measures 72 kilometres on the ground.
A
B
C
D
E
F
G
128°E
130°E
132°E
134°E
136°E
138°E
140°E
12°S
14°S
16°S
18°S
20°S
22°S
24°S
26°S
1
2
3
4
5
6
7
8

South Australia 

coat of arms
South Australian Government

floral emblem
Sturt's desert pea

animal emblem
hairy-nosed wombat

bird emblem
piping shrike (unofficial)

marine emblem
leafy seadragon

mineral/gem emblem
opal

Naracoorte Caves 

Amazing examples of extinct Australian animals have been found in Naracoorte Caves in South Australia. Fossils of bones and teeth belonging to animals that once lived in the area have been discovered. Remains of a marsupial lion, a wombat the size of a hippopotamus and a giant kangaroo have been found at the site. These large Australian animals are called megafauna.

Australian megafauna (page 37)

This skeleton of a marsupial lion has been built from bones found in Naracoorte Caves.

Painting of a marsupial lion

South Australia facts

Capital city: Adelaide, F3

Highest point: Mt Woodroffe, B7 (1435 m)

Longest river: Murray, F3 (700 km within state, 2375 km in total)

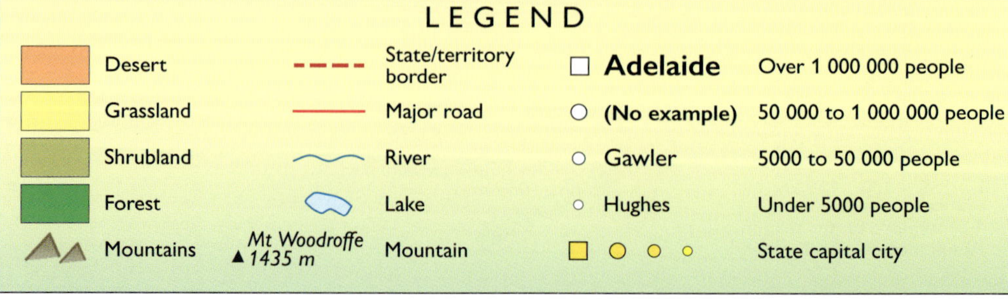

A 130°E B 132°E C 134°E D 136°E E 138°E F 140°E G

Northern Territory

Queensland

Western Australia

New South Wales

Victoria

South Australia

MUSGRAVE RANGES
Mt Everard 1173 m
Ernabella
Mt Woodroffe 1435 m
Great Victoria Desert
Oodnadatta
Warburton Creek
Sturt Stony Desert
Innamincka
Moomba
Cooper Creek
Strzelecki Creek
Kati Thanda-Lake Eyre
Coober Pedy
Marree
Maralinga
Ooldea
Hughes
Cook
Tarcoola
Roxby Downs
Andamooka
Leigh Creek
FLINDERS RANGES
Lake Frome
Lake Torrens
Woomera
NULLARBOR PLAIN
Yalata
St Mary Peak 1165 m
Lake Gairdner
Hawker
Wilson Bluff
Coorabie
Fowlers Bay
Cape Adieu
Ceduna
Cockburn
Smoky Bay
Port Augusta
Iron Knob
Streaky Bay
Mt Remarkable 960 m
Peterborough
Great Australian Bight
Whyalla
Port Pirie
Anxious Bay
Mt Bryan 932 m
Lock
Elliston
Investigator Group
Eyre Peninsula
Burra
Clare
Kadina
Spencer Gulf
Cummins
Murray River
Renmark
Berri
Maitland
Gawler
Yorke Peninsula
Port Lincoln
Gulf St Vincent
Adelaide
West Point
Yorketown
Stenhouse Bay
Murray Bridge
Pinnaroo
Goolwa
Victor Harbor
Lake Alexandrina
Kingscote
Kangaroo Island
INDIAN OCEAN
Bordertown
Kingston South East
Naracoorte
Millicent
Mount Gambier
Cape Northumberland

N
W E
S

0 50 100 150 km
1 centimetre on the map measures 64 kilometres on the ground.

26°S 28°S 30°S 32°S 34°S 36°S 38°S
8 7 6 5 4 3 2 1

Queensland

coat of arms
Queensland Government

floral emblem
Cooktown orchid

animal emblem
koala

bird emblem
brolga

marine emblem
Barrier Reef anemonefish

mineral/gem emblem
sapphire

Coal production

Coal is a fuel that is burned to create heat. The heat energy is then used to make electricity. Queensland is one of the world's largest exporters of coal. Coal is dug out of enormous mines and transported by trains and trucks to nearby ports. The coal is loaded onto special ships and sent overseas.

Producing heat (page 42)

Coal is made from plant remains. Over millions of years, pressure and heat turns this plant material into coal.

Trucks remove coal from the Coppabella mine in Queensland.

Queensland facts

Capital city: Brisbane, D1

Highest point: Mt Bartle Frere, C3 (1611 m)

Longest river: Flinders, B3 (1004 km)

PAPUA NEW GUINEA
Torres Strait
Thursday Island
Cape York
Bamaga
1 centimetre on the map measures 95 kilomtres on the ground.
0 100 200 300 km
Weipa
Lockhart River
Cape York Peninsula
GREAT DIVIDING RANGE
GREAT BARRIER REEF
Princess Charlotte Bay
Gulf of Carpentaria
Pormpuraaw
Cape Flattery
Cooktown
Mitchell River
Mornington Island
Port Douglas
Mareeba
Cairns
Mt Bartle Frere 1611 m
Innisfail
Burketown
Normanton
Gilbert River
Cardwell
Hinchinbrook Island
Georgetown
Halifax Bay
Townsville
Ayr
Bowen
Flinders River
Camooweal
Charters Towers
Whitsunday Island
Lake Dalrymple
Cloncurry
Mount Isa
Hughenden
Mt William 1259 m
Mackay
Queensland
Northern Territory
Georgina River
Winton
Boulia
River
Longreach
Emerald
Yeppoon
Rockhampton
Tropic of Capricorn
Blackwater
Gladstone
Diamantina River
Thomson River
Biloela
Consuelo Peak 1174 m
Bundaberg
Sandy Cape
Hervey Bay
Fraser Island
Maryborough
Windorah
Simpson Desert
Birdsville
Creek
Lake Yamma Yamma
Charleville
Maranoa River
Roma
Gympie
Noosa
Caloundra
Sunshine Coast
Condamine
Dalby
Esk
Toowoomba
Ipswich
Brisbane
South Australia
Cooper
Warrego
Thargomindah
St George
Cunnamulla
Goondiwindi
Warwick
Gold Coast
Dirranbandi
West Barney Peak 1359 m
Kati Thanda-Lake Eyre
New South Wales
CORAL SEA
PACIFIC OCEAN
N
W
E
S
A
B
C
D
1
2
3
4
5
140°E
145°E
150°E
155°E
10°S
15°S
20°S
25°S

New South Wales 

coat of arms	floral emblem	animal emblem	bird emblem	marine emblem
New South Wales Government	*waratah*	*platypus*	*kookaburra*	*blue groper*

Sydney Cove 

Sydney Cove was the site chosen by Captain Arthur Phillip for Australia's first settlement. It has grown to become Sydney, Australia's largest city. The Sydney region was first inhabited by a number of groups of Aboriginal people. Their camps were mainly located close to the shore, where they had a supply of fresh fish and shellfish.

First Fleet (page 58)

New South Wales facts

Capital city: Sydney, F3

Highest point: Mt Kosciuszko, E1 (2228 m)

Longest river: Murray, B2 (1808 km within state, 2375 km in total)

This painting shows the early settlement at Sydney Cove.

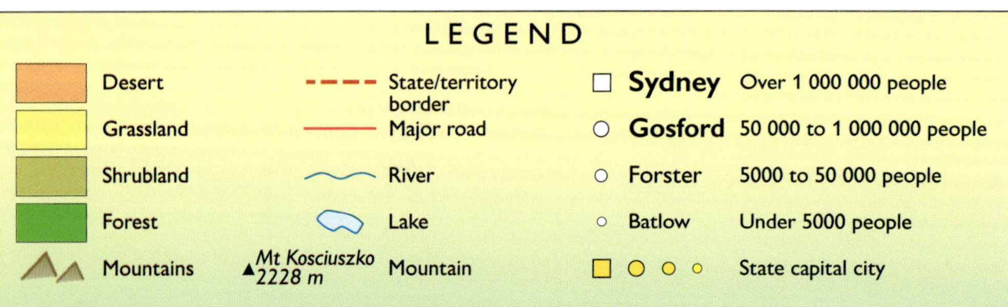

LEGEND					
Desert	State/territory border	**Sydney**	Over 1 000 000 people		
Grassland	Major road	**Gosford**	50 000 to 1 000 000 people		
Shrubland	River	Forster	5000 to 50 000 people		
Forest	Lake	Batlow	Under 5000 people		
Mountains	Mt Kosciuszko 2228 m Mountain		State capital city		

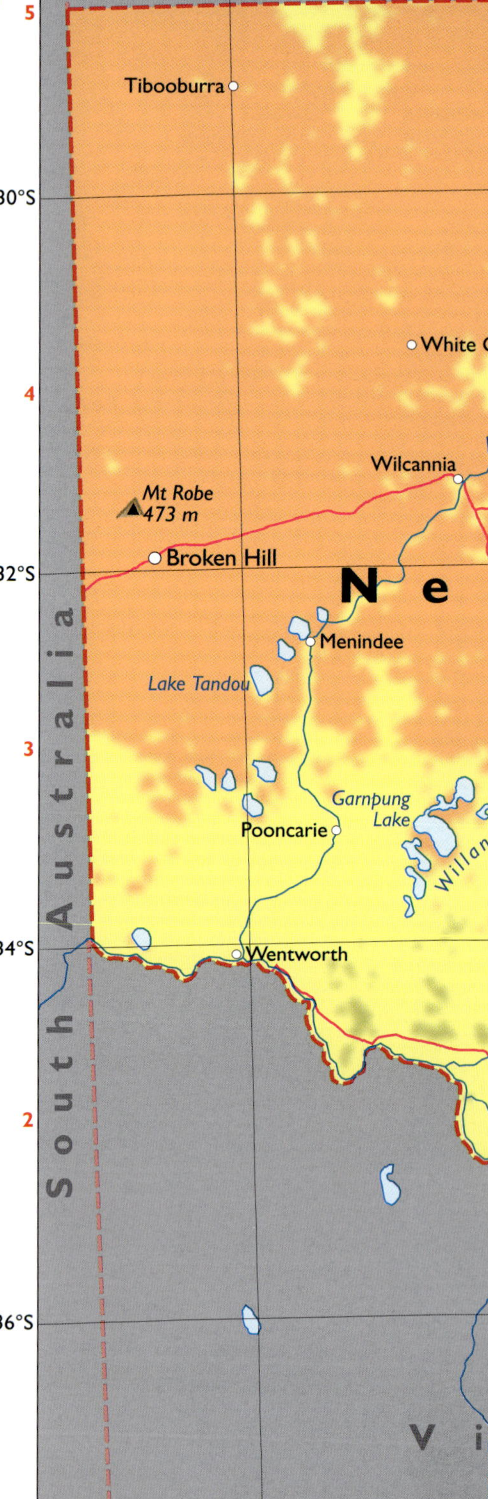

C 146°E D 148°E E 150°E F 152°E G 154°E

28°S 30°S 32°S 34°S 36°S

Queensland

New South Wales

Tweed Heads
Mt Warning 1156 m
Byron Bay
Cape Byron
Lismore
Ballina
Bonshaw
Dumaresq River
Mt Bajimba 1446 m
Clarence River
Woody Head
Moree
Glen Innes
Inverell
Grafton
Walgett
Bourke
Namoi River
Narrabri
Mt Kaputar 1508 m
Round Mountain 1608 m
Armidale
Coffs Harbour
Bogan River
Castlereagh River
Quambone
Coonamble
Nambucca Heads
Gunnedah
WARRUMBUNGLE RANGE
Tamworth
Macleay River
Kempsey
Coonabarabran
Cobar
Black Sugarloaf 1494 m
Port Macquarie
Gilgandra
Babinda Hill 439 m
GREAT DIVIDING RANGE
Taree
Barrington Tops 1555 m
Dubbo
Muswellbrook
Forster
Tallebung
Mudgee
Hunter River
Nelson Bay
Maitland
Cessnock
Newcastle
Lachlan River
Parkes
Lake Cargelligo
Orange
Mt Canobolas 1397 m
Forbes
Bathurst
Wyong
Gosford
Lithgow
BLUE MOUNTAINS
Broken Bay
Windsor
Katoomba
Cowra
Sydney
Botany Bay
Griffith
Young
Hay
Temora
Leeton
Crookwell
Bowral
Wollongong
Cootamundra
Narrandera
Kiama
Murrumbidgee River
Goulburn
Nowra
TASMAN SEA
Gundagai
Wagga Wagga
Jervis Bay
Queanbeyan
Deniliquin
Batlow
Australian Capital Territory
Culcairn
Batemans Bay
Corowa
Albury
Victoria
SNOWY MOUNTAINS
Cooma
Narooma
Jindabyne
Mt Kosciuszko 2228 m
Bega
Snowy River
Eden
Disaster Bay
Cape Howe

N
W E
S

0 50 100 150 km

1 centimetre on the map measures 42 kilometres on the ground.

5 4 3 2 1

C 146°E D 148°E E 150°E F 152°E G 154°E

Australian Capital Territory  

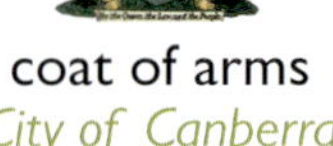

coat of arms
City of Canberra

floral emblem
royal bluebell

bird emblem
gang-gang cockatoo

Australian War Memorial

Located in Canberra, the Australian War Memorial is a national memorial that commemorates the sacrifice of those Australians who have died in war. It includes a shrine, a museum and a large collection of historical military documents. Each year on ANZAC Day (25 April) and Remembrance Day (11 November), the two major days of commemoration in Australia, the Australian War Memorial holds national ceremonies that are attended by thousands of people.

Remembering the ANZACs (page 68)

Red poppies decorate the Roll of Honour at the Australian War Memorial.

Australian War Memorial, Canberra

territory facts
Capital city: Canberra, B2
Highest point: Bimberi Peak, A1 (1912 m)
Longest river: Murrumbidgee, A2 (59 km within territory, 1485 km in total)

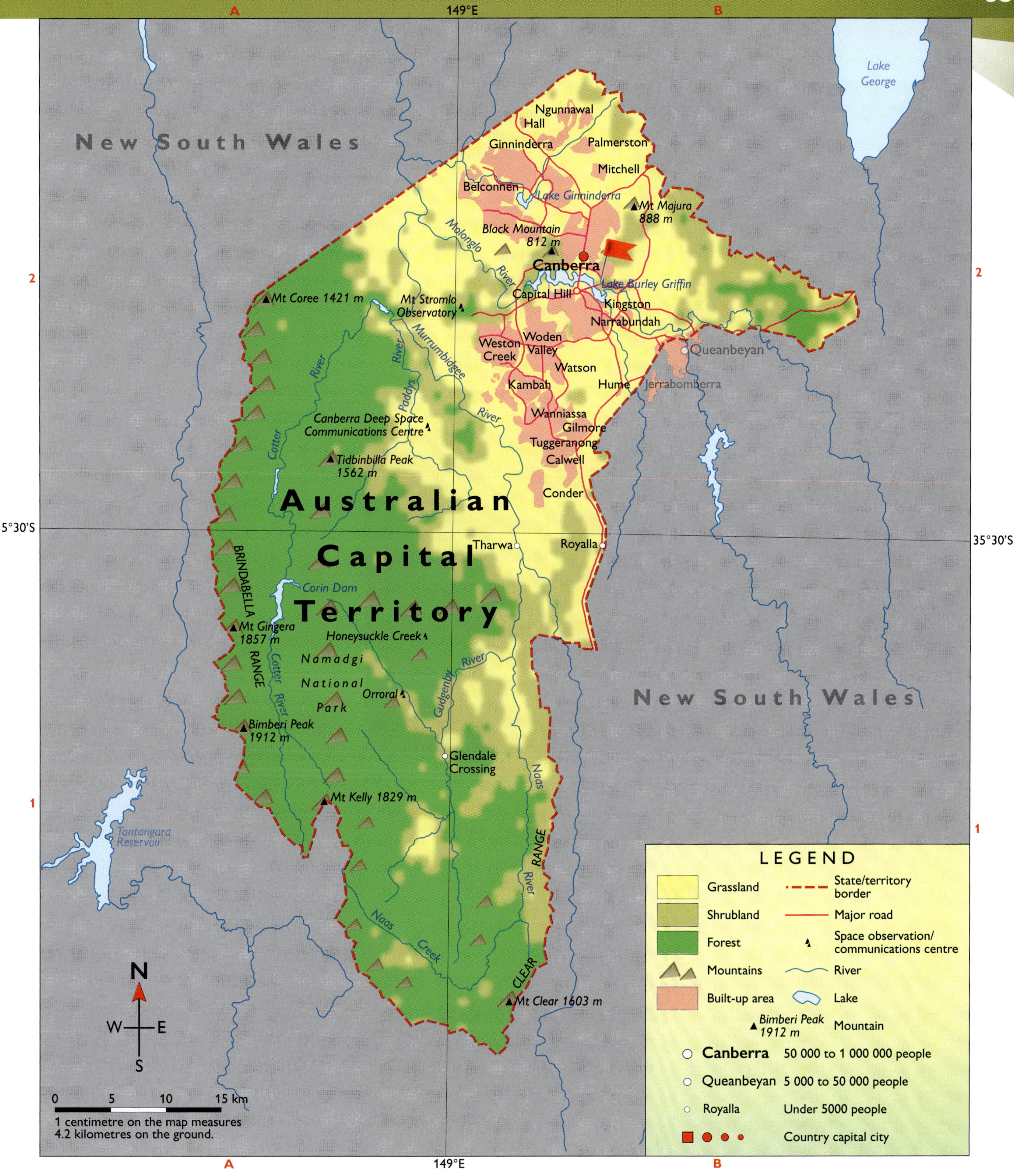
A
149°E
B
New South Wales
Lake George
Ngunnawal
Hall
Ginninderra
Palmerston
Mitchell
Belconnen
Lake Ginninderra
Mt Majura
888 m
Black Mountain
812 m
Molonglo River
Canberra
Lake Burley Griffin
Capital Hill
Kingston
Narrabundah
Mt Coree 1421 m
Mt Stromlo Observatory
Murrumbidgee River
Weston Creek
Woden Valley
Queanbeyan
Watson
Kambah
Hume
Jerrabomberra
Paddys River
Cotter River
Canberra Deep Space Communications Centre
Wanniassa
Gilmore
Tuggeranong
Calwell
Tidbinbilla Peak
1562 m
Conder
Australian Capital Territory
35°30'S
Tharwa
Royalla
BRINDABELLA RANGE
Corin Dam
Mt Gingera
1857 m
Honeysuckle Creek
Namadgi National Park
Orroral
Gudgenby River
New South Wales
Bimberi Peak
1912 m
Glendale Crossing
Naas River
Mt Kelly 1829 m
Tantangara Reservoir
CLEAR RANGE
Naas Creek
Mt Clear 1603 m
2
1
N
W
E
S
0
5
10
15 km
1 centimetre on the map measures
4.2 kilometres on the ground.
LEGEND
Grassland
Shrubland
Forest
Mountains
Built-up area
State/territory border
Major road
Space observation/ communications centre
River
Lake
Bimberi Peak 1912 m
Mountain
Canberra 50 000 to 1 000 000 people
Queanbeyan 5 000 to 50 000 people
Royalla Under 5000 people
Country capital city

Victoria

coat of arms
Victorian Government

floral emblem
common pink heath

animal emblem
Leadbeater's possum

Victoria facts

Capital city: Melbourne, C2

Highest point: Mt Bogong, D2 (1986 m)

Longest river: Goulburn, C2 (654 km)

bird emblem
helmeted honeyeater

marine emblem
weedy seadragon

mineral emblem
gold

A scene from the movie *The Extraordinary Tale of William Buckley*

The amazing William Buckley

On 7 July 1835, early Victorian settlers came across escaped convict William Buckley. He had lived with the native Wathaurong people for 32 years. They taught him their language and how to hunt using their weapons.

Australia's first people (page 52) *First contact* (page 58)

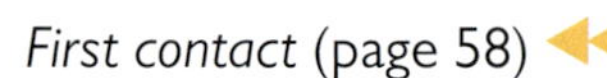

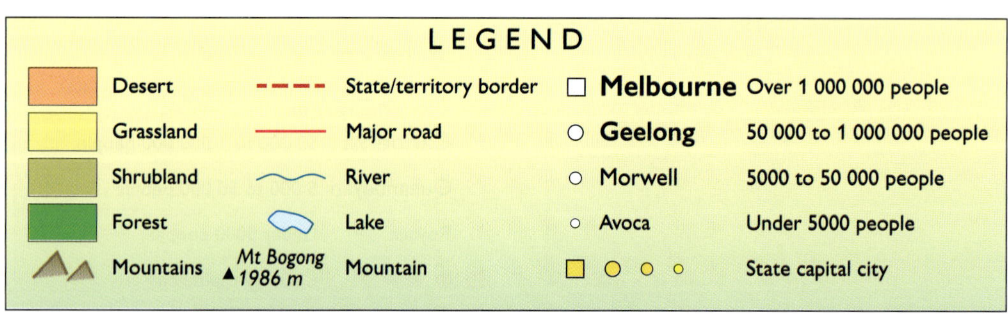

144°E C 146°E D 148°E E 150°E

34°S

3

36°S

2

38°S

1

40°S

New South Wales

Australian Capital Territory

Victoria

GREAT DIVIDING RANGE

SNOWY MOUNTAINS

OTWAY RANGE

Nyah West
Hill
Murray River
Kerang
Cohuna
Kow Swamp
River
River
Charlton
Avoca
Loddon
Wedderburn
Echuca
Goulburn River
Cobram
Yarrawonga
Rutherglen
Ovens
Wodonga
Corryong
Kyabram
Shepparton
Wangaratta
Beechworth
Glenrowan
Benalla
Lake Mokoan
River
Lake Dartmouth
Murchison
Bendigo
Euroa
Mt Buffalo 1723 m
Bright
Mt Bogong 1986 m
Falls Creek
Mt Hotham 1862 m
Dunolly
Heathcote
Maryborough
Avoca
Castlemaine
Seymour
Lake Eildon
Mt Buller 1804 m
Omeo
River
Tambo River
Snowy
Yea
Alexandra
Kilmore
Creswick
Daylesford
Mt Macedon 1011 m
Whittlesea
Marysville
Genoa
Cape Howe
Cann River
Ballarat
Sunbury
Healesville
Yarra River
Skipton
Bacchus Marsh
Melton
Warburton
Bruthen
Orbost
Melbourne
Bairnsdale
Werribee
Emerald
Heyfield
Lake King
Lakes Entrance
Point Hicks
Lake Corangamite
Lara
Berwick
Pakenham
La Trobe River
Lake Wellington
Geelong
Port Phillip Bay
Warragul
Moe
Traralgon
Sale
Winchelsea
Morwell
Ninety Mile Beach
Colac
Torquay
Ocean Grove
Barwon Heads
Sorrento
Cowes
Leongatha
Cape Schanck
Phillip Island
Wonthaggi
Yarram
Lorne
Foster
Apollo Bay
Cape Otway
Cape Liptrap
Wilsons Promontory

King Island

Bass Strait

Flinders Island

Tasmania

144°E C 146°E D 148°E E 150°E F

Tasmania 

coat of arms
Tasmanian Government

floral emblem
Tasmanian blue gum

animal emblem
Tasmanian devil

bird emblem
yellow wattlebird (unofficial)

mineral/gem emblem
crocoite

Tallest flowering plant

Tasmania is home to the world's tallest flowering plant. The 99.8-metre-tall eucalyptus tree grows in a forest south-west of Hobart. The tree, called *Centurion*, is also the world's largest hardwood tree. Its trunk is 405 centimetres in diameter. In the spring, this area is closed to visitors because it is a habitat for eagles. They make their nests high in the branches of trees.

Trees (page 26) *Eggs* (page 38)

A wedge-tailed eagle chick in its nest. Wedge-tailed eagles build their nests in large eucalypt trees so they are sheltered from the wind.

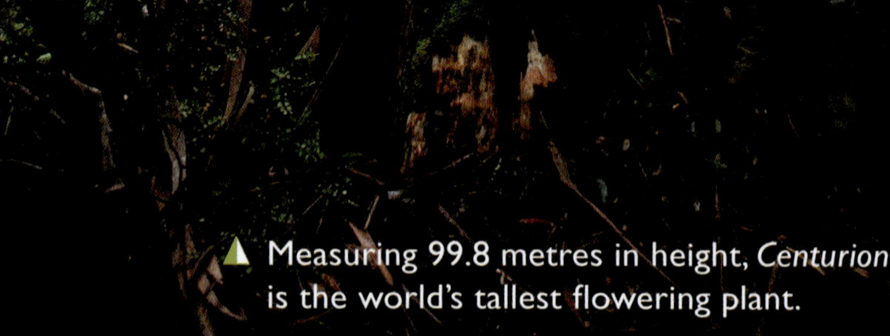

Measuring 99.8 metres in height, *Centurion* is the world's tallest flowering plant.

Tasmania facts

Capital city: Hobart, E2

Highest point: Mt Ossa, D3 (1617 m)

Longest river: South Esk, E3 (245 km)

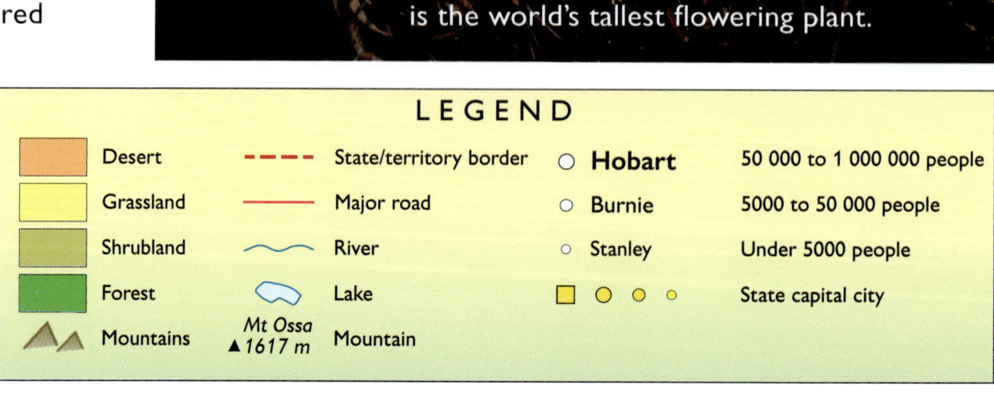

Victoria
Bass Strait
King Island
Currie
Naracoopa
Deal Island
Palana
Flinders Island
Furneaux Group
Whitemark
Lady Barron
Cape Barren Island
Banks Strait
Three Hammock Island
Hunter Island
Robbins Island
Cape Grim
Stanley
Smithton
Marrawah
West Point
Roger River
Wynyard
Burnie
Penguin
Devonport
Beaconsfield
Bell Bay
Noland Bay
Bridport
Gladstone
Cape Portland
Eddystone Point
Derby
St Helens
Mt Barrow 1413 m
Legges Tor 1573 m
St Marys
Fingal
Tamar River
Launceston
Deloraine
Sheffield
Black Bluff 1339 m
Waratah
Savage River
Sandy Cape
Frankland River
Waratah River
Cradle Mountain 1420 m
Barn Bluff 1559 m
Rosebery
Zeehan
Mt Ossa 1617 m
Poatina
Great Lake
Macquarie
South Esk River
Maclean Bay
Bicheno
INDIAN OCEAN
Eldon Peak 1439 m
Tasmania
Ross
Queenstown
Strahan
Cape Sorell
Macquarie Harbour
Derwent Bridge
Franklin River
Frenchmans Cap 1444 m
Tarraleah
Swansea
Coles Bay
Freycinet Peninsula
Oatlands
Triabunna
Gordon River
Derwent River
Mt Field West 1439 m
Hamilton
Maria Island
Bridgewater
Richmond
Sorell
Strathgordon
New Norfolk
Mt Wellington 1269 m
Hobart
Forestier Peninsula
Lake Pedder
Kingston
Eaglehawk Neck
Tasman Peninsula
Port Arthur
Mt Picton 1327 m
Franklin
Cygnet
Storm Bay
North Bruny
Cape Pillar
Dover
Alonnah
Bathurst Harbour
Hastings
South Bruny
South West Cape
South East Cape
TASMAN SEA
N
W
E
S
0 25 50 75 km
1 centimetre on the map measures 23 kilometres on the ground.
A B C D E F
144°E 145°E 146°E 147°E 148°E
39°S 40°S 41°S 42°S 43°S
5 4 3 2 1

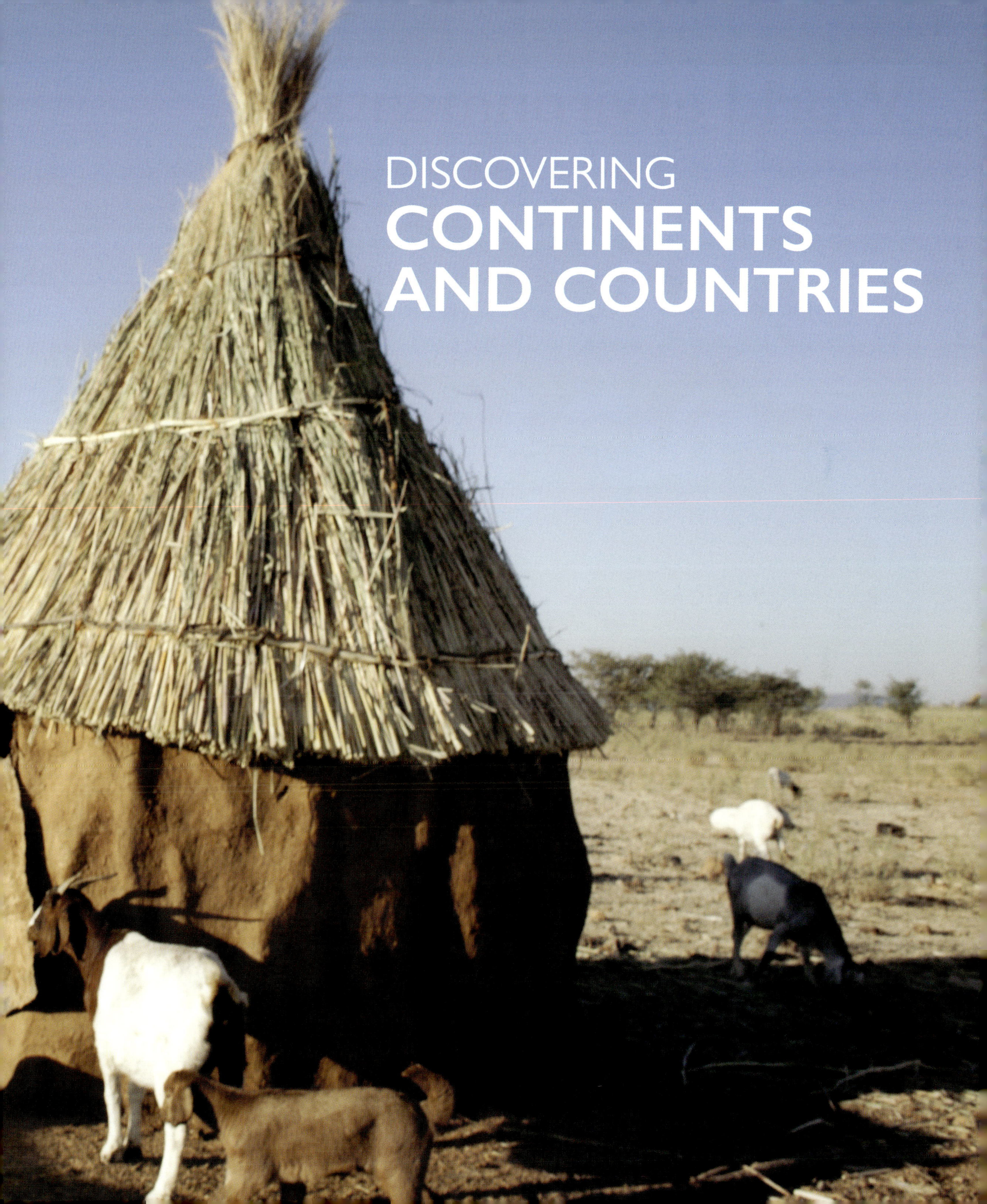

DISCOVERING CONTINENTS AND COUNTRIES

World environments

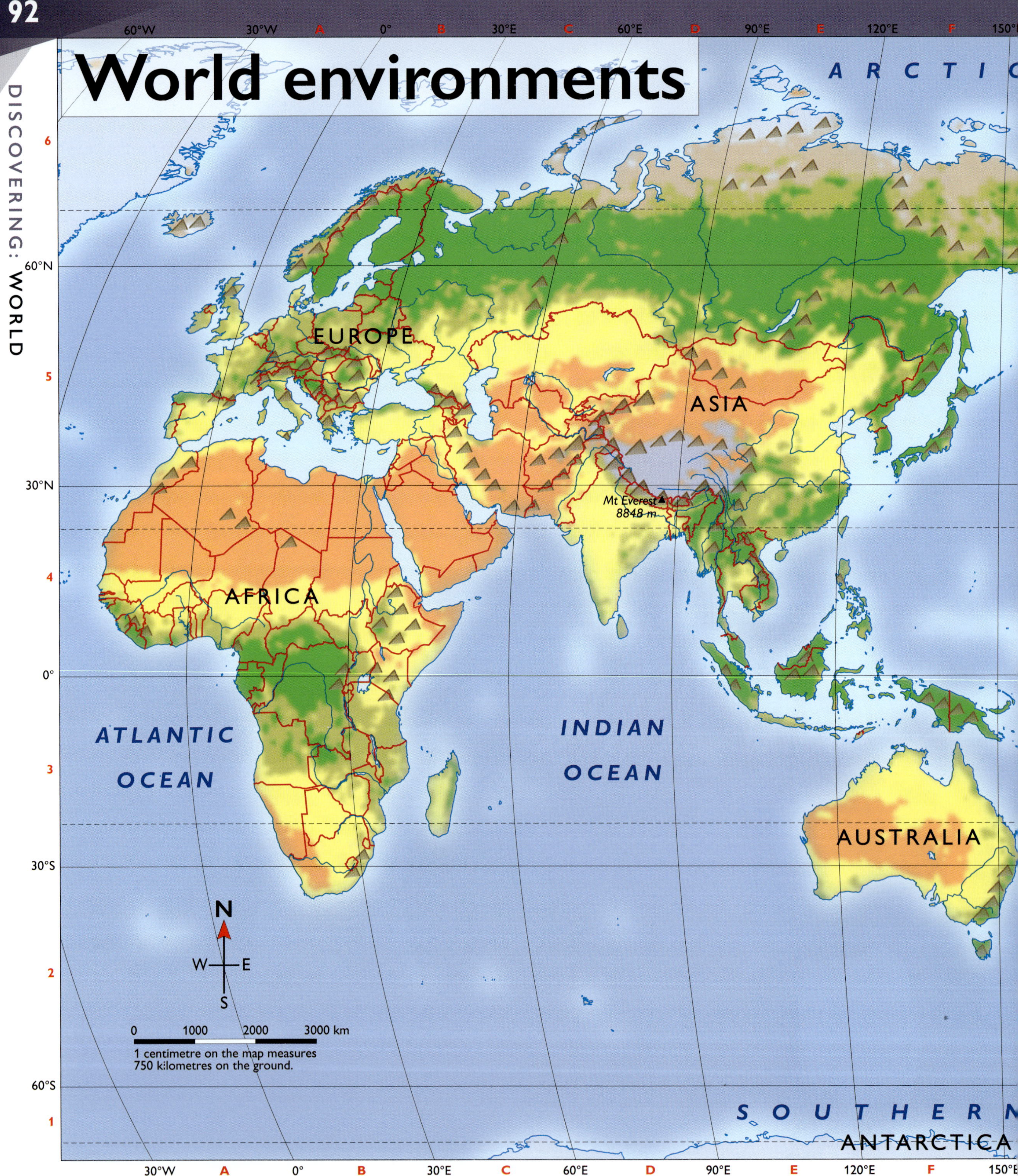

G
180°
H
150°W
I
120°W
J
90°W
K
60°W
L
30°W
M
0°
6
5
4
3
2
1
Arctic Circle
60°N
30°N
Tropic of Cancer
Equator
0°
Tropic of Capricorn
30°S
60°S
Antarctic Circle
O C E A N
NORTH AMERICA
ATLANTIC OCEAN
P A C I F I C
O C E A N
SOUTH AMERICA
O C E A N
LEGEND
Desert
Grassland
Shrubland
Forest
Highland
Tundra
Ice and snow
Mountains
Country border
Mt Everest 8848 m
Highest mountain

World countries

1 **Luxembourg** – Luxembourg
2 **Netherlands** – Amsterdam
3 **Belgium** – Brussels
4 **Switzerland** – Bern
5 **Andorra** – Andorra-la- Vella
6 **Monaco** – Monaco
7 **Liechtenstein** – Vaduz
8 **San Marino** – San Marino
9 **Vatican City**
10 **Czech Republic** – Prague
11 **Slovakia** – Bratislava
12 **Austria** – Vienna
13 **Hungary** – Budapest
14 **Moldova** – Kishinev
15 **Slovenia** – Ljubljana
16 **Croatia** – Zagreb
17 **Bosnia and Herzegovina** – Sarajevo
18 **Montenegro** – Podgorica
19 **Serbia** – Belgrade
20 **Kosovo** – Pristina
21 **North Macedonia** – Skopje
22 **Albania** – Tirane
23 **Malta** – Valletta
24 **Lebanon** – Beirut
25 **Israel** – Jerusalem
26 **Kuwait** – Kuwait
27 **Bahrain** – Manama
28 **Qatar** – Doha
29 **United Arab Emirates** – Abu Dhabi
30 **Senegal** – Dakar
31 **Gambia** – Banjul
32 **Guinea Bissau** – Bissau
33 **Sierra Leone** – Freetown
34 **Ivory Coast** – Yamoussoukro
35 **Ghana** – Accra
36 **Togo** – Lome
37 **Benin** – Porto-Novo
38 **Equatorial Guinea** – Malabo
39 **Sao Tome and Principe** – Sao Tome
40 **Rwanda** – Kigali
41 **Burundi** – Bujumbura
42 **Belize** – Belmopan
43 **Jamaica** – Kingston
44 **St Kitts and Nevis** – Basseterre
45 **Antigua and Barbuda** – St John's
46 **Dominica** – Roseau
47 **St Lucia** – Castries
48 **Barbados** – Bridgetown
49 **St Vincent and the Grenadines** – Kingstown
50 **Grenada** – St George's

G 180° H 150°W I 120°W J 90°W K 60°W L 30°W M 0° N 30°E

OCEAN

Kalaallit Nunaat
(Greenland)
(Denmark)

Arctic Circle

Reykjavik ICELAND

Alaska
(USA)

CANADA

60°N

Ottawa

St Pierre and Miquelon
(France)

ATLANTIC

OCEAN

Azores
(Portugal)

UNITED STATES OF AMERICA

Washington DC

Bermuda (UK)

30°N

PACIFIC

OCEAN

BAHAMAS
Nassau

Tropic of Cancer

Havana CUBA

Hawaii
(USA)

Socorro Island
(Mexico)

MEXICO
Mexico City

Clarion Island
(Mexico)

HAITI
Port-au-Prince

DOMINICAN REPUBLIC
Santo Domingo

CAPE VERDE
Praia

GUATEMALA
Guatemala City

42 43 44 45 46 47 48 49 50

HONDURAS
Tegucigalpa

San Salvador
EL SALVADOR

NICARAGUA
Managua

COSTA RICA
San Jose

PANAMA
Panama City

Caracas

TRINIDAD AND TOBAGO
Port of Spain

GUYANA
Georgetown

Paramaribo

French Guiana (France)

SURINAME

VENEZUELA

Bogota
COLOMBIA

Clipperton
(France)

MARSHALL ISLANDS
Majuro

NAURU
Yaren

South Tarawa

KIRIBATI

Equator 0°

Galapagos Islands
(Ecuador)

Quito
ECUADOR

SOLOMON ISLANDS
Honiara

Funafuti
TUVALU

PERU
Lima

BRAZIL
Brasilia

SAMOA
Apia

American Samoa
(USA)

VANUATU
Port Vila

FIJI
Suva

French Polynesia
(France)

NIUE
Alofi

COOK ISLANDS
Avarua

TONGA
Nukualofa

New Caledonia
(France)

La Paz
BOLIVIA
Sucre

PARAGUAY
Asuncion

Tropic of Capricorn

Pitcairn Island
(UK)

Rapa Nui
(Easter Island)
(Chile)

San Felix
(Chile)

CHILE

Norfolk Island
(Australia)

International Date Line

30°S

Juan Fernandez Islands
(Chile)

Santiago

Buenos Aires
ARGENTINA

URUGUAY
Montevideo

Wellington
NEW ZEALAND

Chatham Islands
(New Zealand)

Auckland Islands
(New Zealand)

LEGEND
Country border
Disputed country border
Country capital city
Azores (Portugal) Dependency

N
W E
S

Falkland Islands (UK)

South Georgia (UK)

6 5 4 3 2 1

0 1000 2000 3000 km

1 centimetre on the map measures 750 kilometres on the ground.

60°S

OCEAN

Antarctic Circle

ANTARCTICA

G 180° H 150°W I 120°W J 90°W K 60°W L 30°W M 0°

Pacific

Pacific Islanders

The Pacific Islands region is made up of very small islands, such as Tahiti, and larger islands, such as Papua New Guinea. The three major groups of people in the region are Melanesians—those who originally inhabited New Guinea, New Caledonia, Torres Strait Islands, Vanuatu, Fiji and the Solomon Islands; Polynesians—those who originally inhabited New Zealand and the eastern Pacific Islands; and Micronesians—those who originally inhabited the north-western Pacific Islands.

Communities around the world (page 62)

Regions of the Pacific

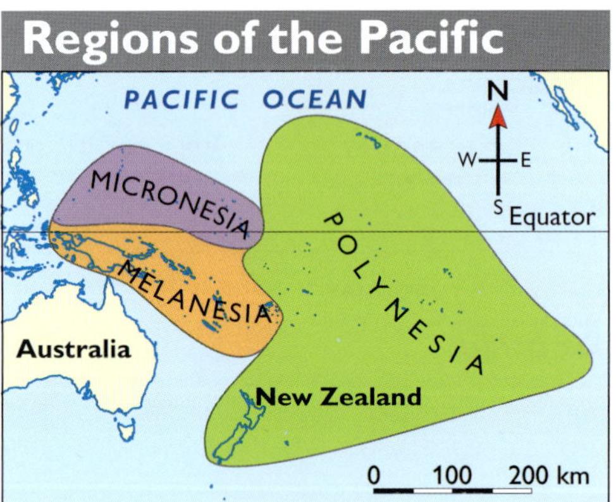

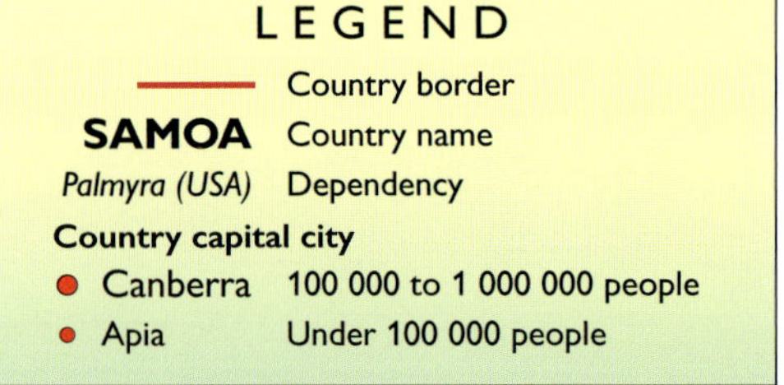
LEGEND

(red line)	Country border
SAMOA	Country name
Palmyra (USA)	Dependency
Country capital city	
● Canberra	100 000 to 1 000 000 people
• Apia	Under 100 000 people

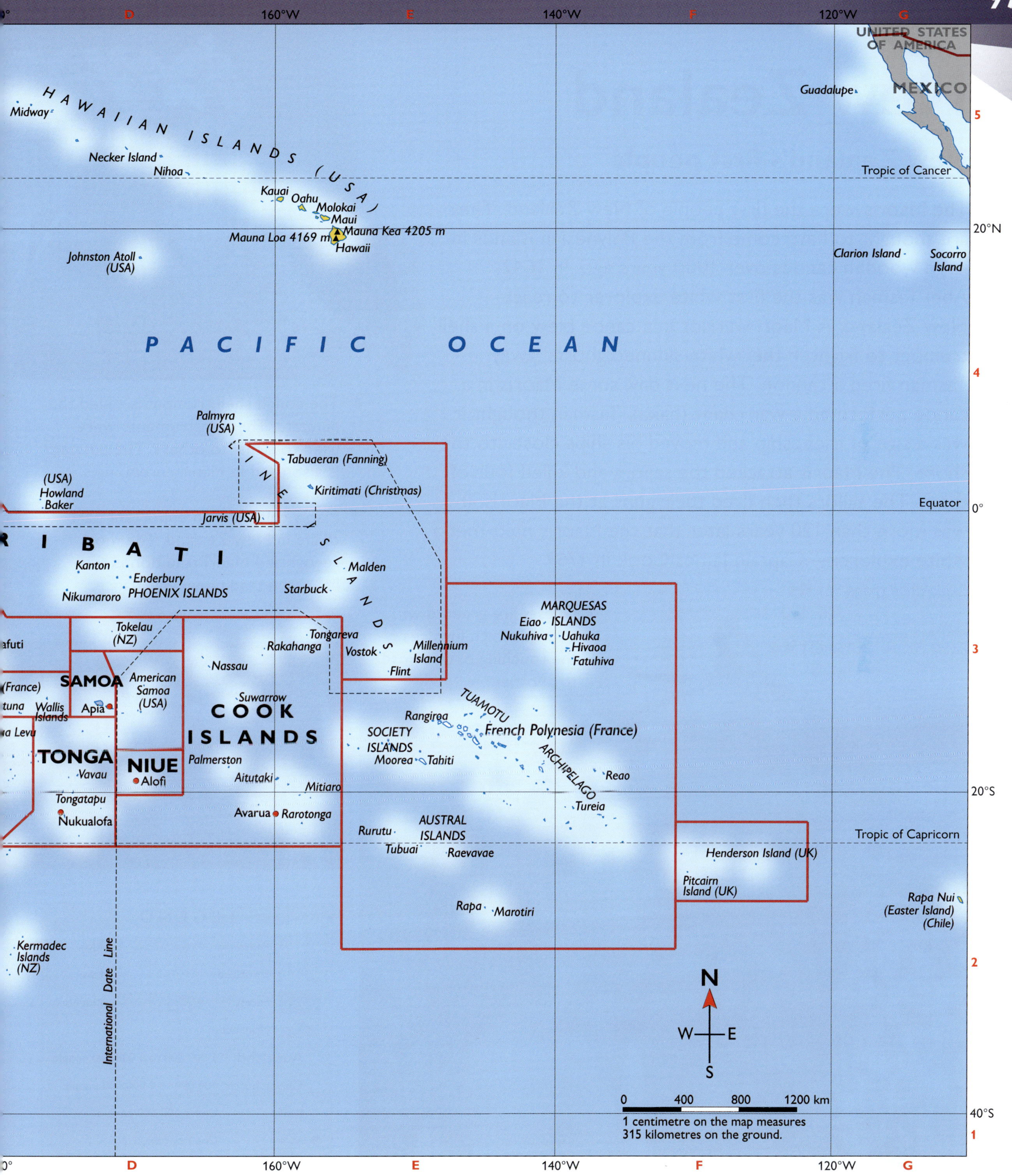
HAWAIIAN ISLANDS (USA)
Midway
Necker Island
Nihoa
Kauai
Oahu
Molokai
Maui
Mauna Loa 4169 m
Mauna Kea 4205 m
Hawaii
Johnston Atoll (USA)
UNITED STATES OF AMERICA
MEXICO
Guadalupe
Tropic of Cancer
Clarion Island
Socorro Island
PACIFIC OCEAN
Palmyra (USA)
LINE ISLANDS
Tabuaeran (Fanning)
Kiritimati (Christmas)
(USA) Howland Baker
Jarvis (USA)
Equator
KIRIBATI
Kanton
Enderbury
Nikumaroro
PHOENIX ISLANDS
Starbuck
Malden
Tokelau (NZ)
Tongareva
Rakahanga
Vostok
Millennium Island
Flint
Nassau
SAMOA
American Samoa (USA)
Apia
Wallis Islands
(France)
Suwarrow
COOK ISLANDS
TONGA
Vavau
NIUE
Alofi
Palmerston
Aitutaki
Mitiaro
Tongatapu
Nukualofa
Avarua
Rarotonga
MARQUESAS ISLANDS
Eiao
Nukuhiva
Uahuka
Hivaoa
Fatuhiva
TUAMOTU ARCHIPELAGO
Rangiroa
French Polynesia (France)
SOCIETY ISLANDS
Moorea
Tahiti
Reao
Tureia
AUSTRAL ISLANDS
Rurutu
Tubuai
Raevavae
Rapa
Marotiri
Tropic of Capricorn
Henderson Island (UK)
Pitcairn Island (UK)
Rapa Nui (Easter Island) (Chile)
Kermadec Islands (NZ)
International Date Line
N
W
E
S
0
400
800
1200 km
1 centimetre on the map measures 315 kilometres on the ground.
160°W
140°W
120°W
20°N
0°
20°S
40°S
D
E
F
G
5
4
3
2
1

New Zealand

New Zealand's first people

The Māoris were the first people of New Zealand. They are thought to have come from the Polynesian islands in large wooden canoes over 1000 years ago. In 1642, Abel Tasman was the first white explorer to reach New Zealand. A Māori warrior in a canoe blew on a shell trumpet to frighten the 'white skinned ghosts' away. Tasman fired a cannon. The next day, some Māoris in a canoe performed a war chant (haka). Tasman thought it was a sign of friendship and moved his ships closer to the shore. The Māoris attacked the sailors and killed four of them. The sailors then fired their guns at the Māoris. It was more than 120 years later that the Māoris next met a white explorer—Captain James Cook in 1769.

Exploration (page 56)
First contact (page 58)
Living cultures (page 61)

The conch shell trumpet is called the *putatara*. These instruments were used to signal to others. The *putatara* is also used in Māori ceremonies.

Māori culture is an important part of life in New Zealand. The most recognised Māori tradition is the haka, the traditional war dance.

People and culture (page 62)

Guests are greeted with a *hongi*, which involves the touching of one nose to another.

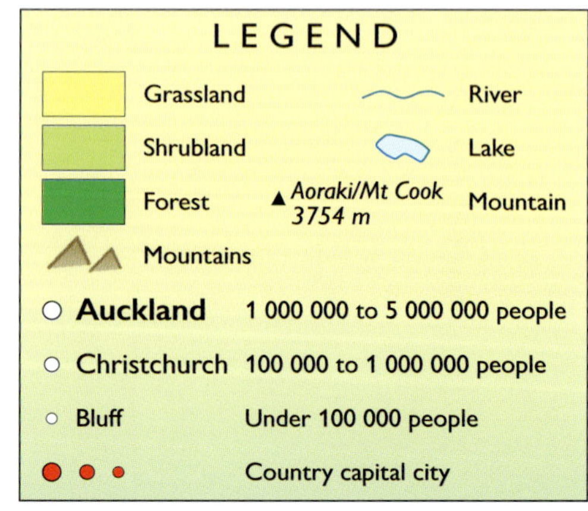

NEW ZEALAND
North Island
South Island
TASMAN SEA
PACIFIC OCEAN
North Cape
Tauroa Point
Kaitaia
Bay of Islands
Waitangi
Northland
Dargaville
Great Barrier Island
Wellsford
Kaipara Harbour
Coromandel Peninsula
North Shore
Waitakere
Whitianga
Auckland
Manukau
Waihi
Tauranga
Bay of Plenty
Hamilton
East Cape
Whakatane
Hikurangi 1752 m
Rotorua
Te Kuiti
RAUKUMARA RANGE
Rangitaiki River
Taupo
Lake Taupo
Gisborne
Taumarunui
New Plymouth
Mt Taranaki (Egmont) 2518 m
Cape Egmont
Mt Ruapehu 2797 m
Mahia Peninsula
Hawke Bay
Wanganui River
Napier
Hastings
RUAHINE RANGE
Wanganui
Waipukurau
Palmerston North
Foxton
Cape Farewell
Tasman Bay
Paraparaumu
Cook Strait
Nelson
Picton
Wellington
Blenheim
Cape Palliser
Cape Foulwind
Westport
SPENSER MOUNTAINS
Tapuaenuku 2885 m
Greymouth
Kaikoura
Hokitika
Lake Coleridge
Pegasus Bay
SOUTHERN ALPS
Franz Josef Glacier
Mt Tasman 3498 m
Darfield
Christchurch
Aoraki/Mt Cook 3754 m
Methven
Canterbury Plains
Ashburton
Canterbury Bight
Lake Pukaki
Mt Aspiring 3027 m
Timaru
Lake Hawea
Milford Sound
Lake Wanaka
Wanaka
Waitaki River
Caswell Sound
Queenstown
Oamaru
Alexandra
Lake Te Anau
Clutha River
Otago Peninsula
Fiordland
Dunedin
Winton
Milton
Puysegur Point
Invercargill
Foveaux Strait
Bluff
Waipapa Point
Stewart Island
N
W
E
S
0
50
100
150 km
1 centimetre on the map measures 60 kilometres on the ground.
170°E
174°E
178°E
166°E
36°S
40°S
44°S
A
B
C
D
E
1
2
3
4

Asia

Asia is the most populated continent. About 60 per cent of the world's people live here. It is made up of 48 countries whose people speak several thousand different languages. Asia is home to some of the world's major religions, such as Buddhism, Hinduism and Islam.

Asia facts

Population: 4 377 000 000

Largest city: Tokyo 38 140 000 people

World's tallest statue

◄ The Spring Temple Buddha in China is 153 metres tall. It is the tallest statue in the world. Buddhism is a religion that began in India, but is now practised throughout the world.

Different religions (page 67) ◄◄

ARCTIC OCEAN
BERING SEA
KARA SEA
LAPTEV SEA
Arctic Circle
SEA OF OKHOTSK
RUSSIA
Lake Baikal
Ulan Bator
MONGOLIA
NORTH KOREA
Pyongyang
Seoul
SOUTH KOREA
SEA OF JAPAN
Tokyo
JAPAN
Beijing
CHINA
YELLOW SEA
Yangtze River
EAST CHINA SEA
Tropic of Cancer
Taipei
TAIWAN
PACIFIC OCEAN
Thimphu
BHUTAN
BANGLADESH
Dhaka
MYANMAR
Naypyidaw
Hanoi
LAOS
Vientiane
VIETNAM
PHILIPPINES
Manila
Yangon
THAILAND
SOUTH CHINA SEA
Bangkok
CAMBODIA
Phnom Penh
ANDAMAN SEA
Bandar Seri Begawan
BRUNEI
Equator
MALAYSIA
Kuala Lumpur
SINGAPORE
Singapore
INDONESIA
ARAFURA SEA
Dili
TIMOR-LESTE
Jakarta
N
W
E
S

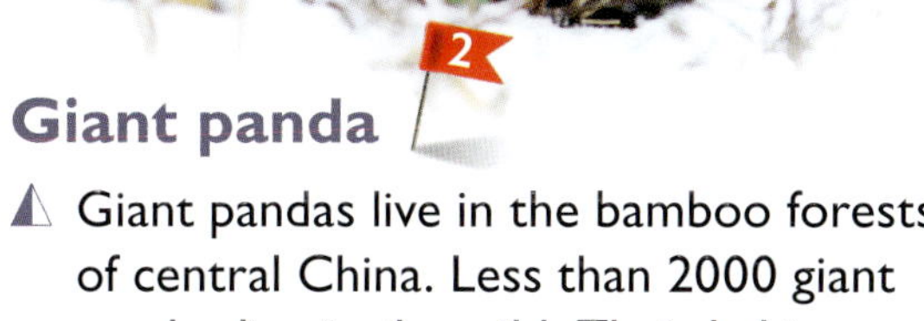

Giant panda

▲ Giant pandas live in the bamboo forests of central China. Less than 2000 giant pandas live in the wild. Their habitat has been cleared for farms.

Deforestation (page 26) ◀◀

Habitats (page 28) ◀◀

World's largest gold mine

▼ The Grasberg Mine, located in the mountains of Papua in Indonesia, is the largest gold mine in the world. The mine includes a huge open-cut pit as well as an underground mine.

Deforestation (page 26) ◀◀

Gold in Australia (page 31) ◀◀

South-East Asia

Halong Bay

The coastline of Halong Bay in Vietnam has been eroded into amazing shapes. The soft limestone has formed thousands of small islands. Many of the islands are hollow with huge caves.

The changing coastline (page 24)

Halong Bay in Vietnam

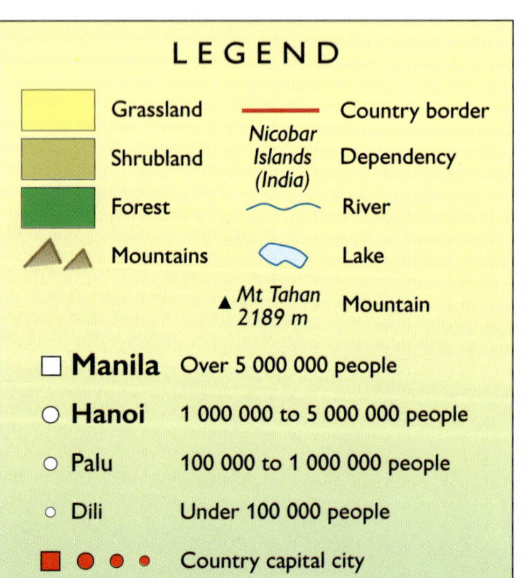

INDIA
BANGLADESH
CHINA
MYANMAR
THAILAND
LAOS
VIETNAM
CAMBODIA
MALAYSIA
SINGAPORE
INDIAN OCEAN
ANDAMAN SEA
SOUTH CHINA SEA
JAVA SEA
Bay of Bengal
Gulf of Tonkin
Gulf of Thailand
Strait of Melaka
Chittagong
Mandalay
Meiktila
Taunggyi
Sittwe
Naypyidaw
Chiang Mai
Henzada
Pegu
Yangon
Moulmein
Cape Negrais
Andaman Islands (India)
Port Blair
Mergui
Lanbi Island
Isthmus of Kra
Nicobar Islands (India)
Surat Thani
Phuket
Songkhla
Banda Aceh
Alor Setar
Penang
George Town
Kota Bharu
Kuala Terengganu
Taiping
Mt Tahan 2189 m
Ipoh
Medan
Leuser 3380 m
Kuantan
Kelang
Kuala Lumpur
Ampang Jaya
Melaka
Kluang
Dumai
Johor Baharu
Singapore
Simeulue
Nias
Pakanbaru
Sumatra
Padang
Siberut
Kerinci 3805 m
Jambi
Bengkulu
Palembang
Bangka
Pangkalpinang
Belitung
Bandar Lampung
Jakarta
Bekasi
Bogor
Bandung
Cirebon
Tasikmalaya
Semarang
Surakarta
Yogyakarta
Surabaya
Probolinggo
Malang
Java
Gejiu
Fan si pan 3414 m
Red River
Nanning
Guangzhou
Macau
Hanoi
Haiphong
Zhanjiang
Haikou
Hainan
Luang Prabang
Thanh Hoa
Lampang
Vientiane
Mekong River
Savannakhet
Hue
Khon Kaen
Da Nang
Bangkok
Siem Reap
Qui Nhon
Rayong
Nha Trang
Phnom Penh
Ho Chi Minh City
Rach Gia
Soc Trang
Ca Mau
Point Baibung
Natuna
Anambas Islands
South Natuna
Cape Datu
Sarawak
Sibu
Kuching
Kapuas River
Pontianak
Raya 2278 m
Kalimantan
Irrawaddy River
Salween River
Christmas Island (Australia)
Cocos Islands (Australia)
0 200 400 600 800 km
1 centimetre on the map measures 200 kilometres on the ground.
N W E S
A B
100°E 110°E
20°N 10°N 0° 10°S
5 4 3 2 1

C 120°E D 130°E E 140°E F

Tropic of Cancer

Xiamen
Shantou
Shenzhen
Hong Kong
TAIWAN
Tainan
Kaohsiung
Taitung
Bashi Channel
Luzon Strait
Cape Engano
Aparri
Luzon
Baguio
Santiago
Dagupan
Mt Pinatubo 1660 m
Manila
Naga
Mindoro
Mindoro Strait
Samar
Roxas
Panay
Tacloban
Leyte
Bacolod
Cebu
Negros
Palawan
MINDANAO SEA
Butuan
Balabac
Balabac Strait
SULU SEA
Marawi
Mindanao
Zamboanga
Moro Gulf
Mt Apo 2954 m
Davao
General Santos
Tinaca Point
Kota Kinabalu
Kinabalu 4101 m
Sandakan
Sabah
Bandar Seri Begawan
BRUNEI
Tawau
PHILIPPINES

PACIFIC OCEAN

NORTHERN MARIANAS
Saipan
Guam (USA)
PHILIPPINE SEA
FEDERATED STATES OF MICRONESIA
Babelthuap
Melekeok
PALAU

Talaud Islands
SULAWESI SEA
Sangihe Islands
Morotai
Halmahera
MALUKU SEA
Borneo
Menyapa 2000 m
Manado
Gorontalo
Minahassa Peninsula
Makassar Strait
Samarinda
Tomini Bay
Penju Islands
Palu
Balikpapan
Lake Poso
Sula Islands
Obi
Waigeo
Kwoka 3000 m
Yapen
Misool
Gulf of Cenderawasih
New Guinea
Jayapura
VANREES RANGE
Mamberamo River
Sulawesi
Banggai Islands
Taliabu
SERAM SEA
Seram
Buru
MAOKE RANGE
Banjarmasin
Rantekombola 3455 m
Jaya Peak 5040 m
Mandala 4703 m
Parepare
Kendari
Bone Bay
Cape Selatan
Makassar
Butung
Papua
PAPUA NEW GUINEA
Kai Islands
BANDA SEA
Aru Islands
Salayar
INDONESIA
Kangean Islands
Damar
FLORES SEA
Wetar
Tanimbar Islands
Dolak
Bali
Lombok
Tambora 2851 m
Lembata
Alor
Dili
Mataram
Flores
Pantar
TIMOR-LESTE
Denpasar
Sumbawa
SAVU SEA
Timor
ARAFURA SEA
Sumba
Kupang
TIMOR SEA
Roti
AUSTRALIA

20°N
10°N
Equator
0°
10°S

5
4
3
2
1

C 120°E D 130°E E 140°E F

East Asia

Making cars in Asia

Japan and China are two of the largest car producers in the world. Many major car brands, such as Toyota, Honda, Suzuki, Mazda and Subaru, are from Japan. Both China and Japan import iron ore from Australia to make steel for the cars. China and Japan export finished cars to Australia.

How is steel made? (page 32)

Robots weld the frame of a car on an assembly line in Japan.

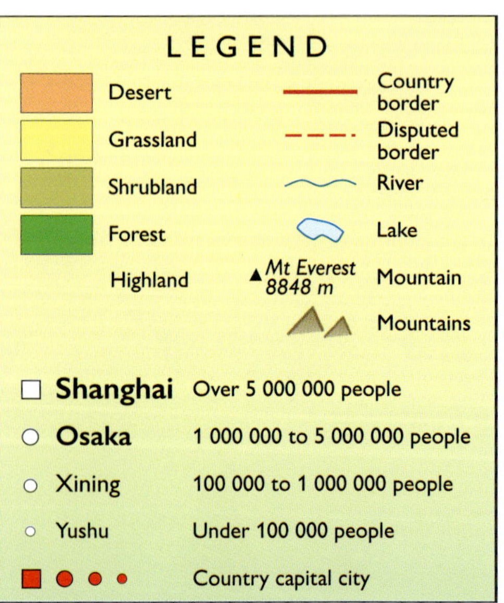

KAZAKHSTAN
KYRGYZSTAN
TAJIKISTAN
AFGHANISTAN
PAKISTAN
INDIA
NEPAL
BHUTAN
BANGLADESH
MYANMAR
THAILAND
CHINA
MONGOLIA
RUSSIA

Astana, Barnaul, Novokuznetsk, Semipalatinsk, Balkhash, Taraz, Bishkek, Alma-Ata, Yining, Shihezi, Urumqi, Namangan, Kashi, Hotan, Jiayuguan, Golmud, Yushu, Srinagar, Amritsar, Ludhiana, Chandigarh, Meerut, New Delhi, Agra, Kanpur, Lucknow, Pokhara, Kathmandu, Xigaze, Lhasa, Thimphu, Itanagar, Gauhati, Varanasi, Patna, Bhopal, Jabalpur, Dhanbad, Ranchi, Asansol, Jamshedpur, Dhaka, Kolkata, Khulna, Nagpur, Bhilai, Chittagong, Myitkyina, Baoshan, Mandalay, Meiktila, Taunggyi, Bhubaneswar, Sittwe, Vishakhapatnam, Vijayawada, Naypyidaw, Chiang Mai, Lampang, Henzada, Pegu

Grandioznyy 2922 m, Mt Belukha 4503 m, Pobeda Peak 7439 m, Muztagh Ata 7546 m, Godwin Austen (K2) 8611 m, Ulugh Muztagh 7726 m, Annapurna 8078 m, Mt Everest 8848 m

ALTAI MOUNTAINS, KHANGAI, KUNLUN MOUNTAINS, HIMALAYAS, TURFAN DEPRESSION, TARIM BASIN, PLATEAU OF TIBET (XIZANG PLATEAU), DECCAN PLATEAU

Lake Balkhash, Lake Zaisan, Lake Issyk, Ubsu Lake, Nam Lake, Irtysh River, Yarkant River, Indus River, Ganges River, Brahmaputra River, Salween River, Mekong River, Irrawaddy River

Bay of Bengal

50°N, 40°N, 30°N, 20°N, 70°E, 80°E, 90°E, 100°E, A, B, C, D, 1, 2, 3, 4

SEA OF OKHOTSK
Sakhalin
Tatar Strait
Yuzhno-Sakhalinsk
Cape Aniva
Kunashir
Cape Soya
Asahikawa
Kushiro
Asahi Peak 2290 m
Hokkaido
Sapporo
Cape Shiriya
Hachinohe
Honshu
Sendai
Niigata
Nagaoka
Cape Rokugo
Mt Shirane 2578 m
Tokyo
Yokohama
Mt Fuji 3776 m
Shizuoka
Hamamatsu
Nagoya
Kyoto
Kobe
Osaka
JAPAN
Okayama
Hiroshima
Cape Shiono
Shikoku
Kitakyushu
Fukuoka
Oita
Kumamoto
Kyushu
SEA OF JAPAN (EAST SEA)
Korea Strait
SIKHOTE MOUNTAINS
Mt Medvezhya 1556 m
Mt Vysokaya 1746 m
Mt Oblachnaya 1855 m
Khabarovsk
Birobidzhan
Vladivostok
Lake Khanka
Jixi
Amur River
Shilka
MANCHURIAN PLAIN
Yichun
Qiqihar
Harbin
Songhua River
Nen River
Mudanjiang
Jilin
Changchun
Shenyang
Anshan
Chongjin
NORTH KOREA
Pyongyang
Nampo
Dalian
Seoul
Incheon
SOUTH KOREA
Daejeon
Daegu
Busan
Gwangju
Jeju Island
Gulf of Chihli
Yantai
Qingdao
YELLOW SEA
Lake Baikal
Munku Sardyk 3491 m
Irkutsk
Ulan-Ude
Chita
Lake Hovsgol
Asaralta 2776 m
Ulan Bator
Kerulen River
Hulun Lake
MOUNTAINS
MONGOLIA
Gobi Desert
Xilin Hot
Baotou
Hohhot
Zhangjiakou
Beijing
Datong
Tianjin
Hwang River
Ordos Desert
Yinchuan
Shijiazhuang
Taiyuan
Handan
Jinan
Xining
Lanzhou
CHINA
Xianyang
Xi'an
Luoyang
Zhengzhou
Zaozhuang
Yancheng
Gaoyou Lake
Nanjing
Hefei
Suzhou
Shanghai
DABA MOUNTAINS
Han River
Chengdu
Yichang
Wuhan
Hangzhou
Ningbo
EAST CHINA SEA
Gongga Mountain 7590 m
Chongqing
Nanchang
Changsha
Fuzhou
Wenzhou
Yangtze River
Gan River
Wu River
Guiyang
Shaoguan
Okinawa
PACIFIC OCEAN
Tropic of Cancer
Taipei
Taichung
TAIWAN
Taitung
Tainan
Kaohsiung
Xiamen
Chaozhou
Shantou
Kunming
Liuzhou
Gejiu
Nanning
Xi River
Guangzhou
Jiangmen
Shenzhen
Macau
Hong Kong
Bashi Channel
Fan si pan 3414 m
Hanoi
Haiphong
Zhanjiang
Haikou
Hainan
Luang Prabang
Thanh Hoa
VIETNAM
LAOS
Vientiane
Gulf of Tonkin
SOUTH CHINA SEA
Luzon Strait
Cape Engano
Aparri
Luzon
PHILIPPINES
PHILIPPINE SEA
N
W
E
S
0 200 400 600 km
1 centimetre on the map measures 170 kilometres on the ground.
110°E
120°E
130°E
140°E
150°E
40°N
30°N
20°N

South Asia

The Ganges River

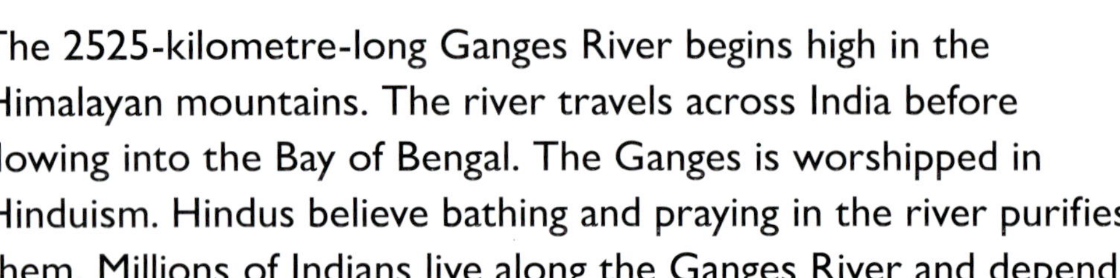

The 2525-kilometre-long Ganges River begins high in the Himalayan mountains. The river travels across India before flowing into the Bay of Bengal. The Ganges is worshipped in Hinduism. Hindus believe bathing and praying in the river purifies them. Millions of Indians live along the Ganges River and depend on it for their daily needs.

Rivers (page 22)

Tens of thousands of Hindus gather for a festival in Patna, India. They bathe and pray in the holy Ganges River. The festival is also an opportunity to have fun with friends and family.

Celebrations (page 67)

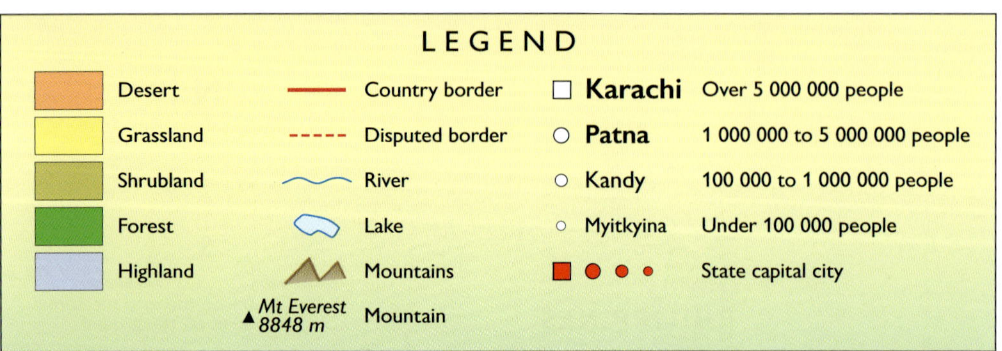

TAJIKISTAN
Dushanbe
Muztagh Ata 7546 m
TARIM BASIN
Hotan
Godwin Austen (K2) 8611 m
KUNLUN MOUNTAINS
Ulugh Muztagh 7726 m
Golmud
Xining
Lanzhou
Tirich Mir 7699 m
HINDU KUSH
KARAKORAM RANGE
Khyber Pass
Peshawar
Islamabad
Rawalpindi
Srinagar
CHINA
PLATEAU OF TIBET
(XIZANG PLATEAU)
Yushu
Lahore
Gujranwala
Amritsar
Faisalabad
Ludhiana
Chandigarh
Multan
HIMALAYAS
Nam Lake
Gongga Mountain 7590 m
Yangtze River
Lhasa
Xigaze
Indus River
Sukkur
Thar Desert
Bikaner
New Delhi
Meerut
Bareilly
Aligarh
Annapurna 8078 m
Pokhara
NEPAL
Kathmandu
Mt Everest 8848 m
Thimphu
BHUTAN
Itanagar
NAGA HILLS
Mekong
Brahmaputra River
Gauhati
Myitkyina
Baoshan
Jaipur
Agra
Lucknow
Ganges
Kanpur
Jodhpur
Gwalior
Patna
Hyderabad
Kotah
Allahabad
Varanasi
BANGLADESH
Dhaka
INDIA
MYANMAR
Dhanbad
Asansol
Ranchi
Jamshedpur
Ahmadabad
Bhopal
Jabalpur
Indore
Vadodara
Rajkot
Kolkata
Khulna
Chittagong
Mandalay
Meiktila
Taunggyi
Salween River
LAOS
Surat
Nagpur
Amravati
Bhilai
Daman
DECCAN PLATEAU
Bhubaneswar
Sittwe
Irrawaddy River
Naypyidaw
Nasik
Aurangabad
PEGU RANGE
Chiang Mai
Lampang
Mumbai
Godavari River
Bay of Bengal
Pune
Nizamabad
Henzada
THAILAND
Sholapur
Hyderabad
Vishakhapatnam
Yangon
Pegu
Moulmein
Kolhapur
Krishna River
Vijayawada
Cape Negrais
Panaji
Hubli
Davangere
Bangkok
Bangalore
Chennai
Andaman Islands (India)
ANDAMAN SEA
Mangalore
Mergui
Port Blair
Lanbi Island
Laccadive Islands
Coimbatore
Tiruchchirappalli
Isthmus of Kra
Kochi
Madurai
Jaffna
Surat Thani
Tiruvananthapuram
Gulf of Mannar
SRI LANKA
Phuket
Cape Comorin
Nicobar Islands (India)
Strait of Melaka
Kandy
Colombo
Pidurutalagala 2524 m
MALDIVES
Dondra Head
Banda Aceh
Sumatra
Medan
INDONESIA
Ari Atoll
Male
70°E
80°E
90°E
100°E
30°N
20°N
10°N
C
D
E
F
1
2
3
4

Middle East

World's tallest building

Burj Khalifa in Dubai is the world's tallest building. Opened in 2010, the 830-metre skyscraper has more than 160 storeys and can be seen from nearly 100 kilometres away. It is nearly 200 metres taller than the world's second-tallest building, Shanghai Tower.

How is steel made? (page 32)

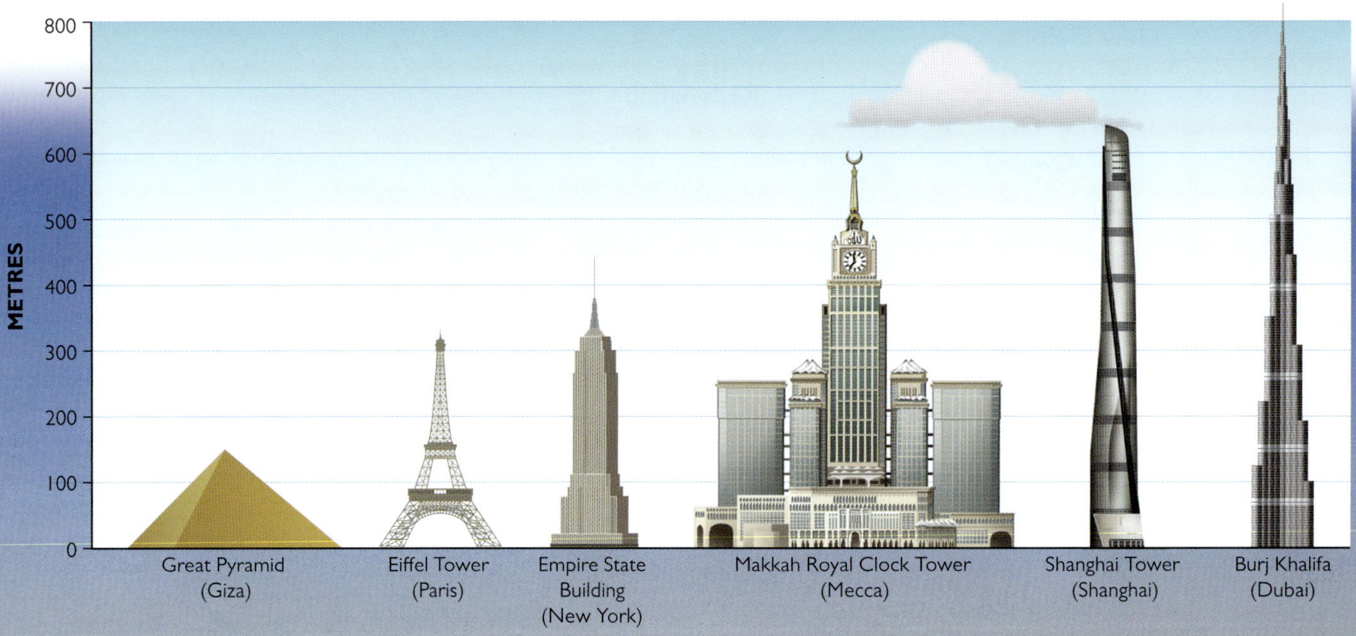

METRES

800
700
600
500
400
300
200
100
0

Great Pyramid (Giza)
Eiffel Tower (Paris)
Empire State Building (New York)
Makkah Royal Clock Tower (Mecca)
Shanghai Tower (Shanghai)
Burj Khalifa (Dubai)

This graph shows the world's tallest three buildings compared to other well-known buildings.

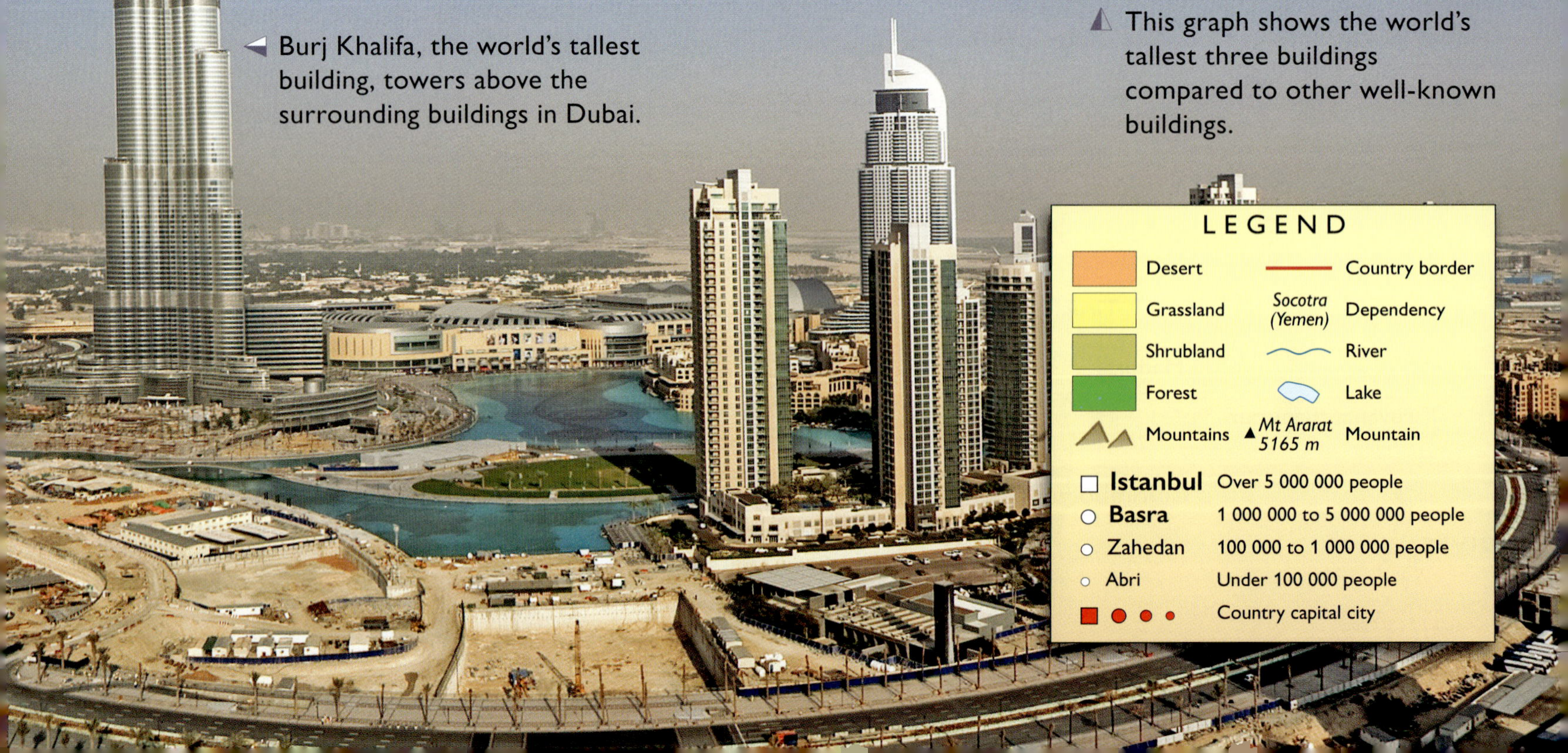

Burj Khalifa, the world's tallest building, towers above the surrounding buildings in Dubai.

LEGEND

	Desert		Country border
	Grassland	*Socotra (Yemen)*	Dependency
	Shrubland		River
	Forest		Lake
	Mountains	▲ *Mt Ararat 5165 m*	Mountain
□	**Istanbul**	Over 5 000 000 people	
○	**Basra**	1 000 000 to 5 000 000 people	
○	Zahedan	100 000 to 1 000 000 people	
○	Abri	Under 100 000 people	
■ ● ● ●		Country capital city	

MOLDOVA
Iasi
Odesa
UKRAINE
Rostov
Volga River
Guryev
KAZAKHSTAN
ROMANIA
Sea of Azov
Astrakhan
Bucharest
Simferopol'
Kerch
RUSSIA
Aral Sea (former extent)
Constanta
Sevastopol'
Krasnodar
Stavropol
BULGARIA
Mt Elbrus 5642 m
Nalchik
UST URT PLATEAU
UZBEKISTAN
Burgas
BLACK SEA
Sochi
Grozny
Aktau
Nukus
Kyzyl Desert
Sukhumi
CAUCASUS MOUNTAINS
Makhachkala
GEORGIA
Kutaisi
CASPIAN SEA
Dashoguz
Istanbul
Zonguldak
Batumi
Tbilisi
Corlu
Izmit
Samsun
Ordu
Bukhara
Bursa
Corum
Kumairi
AZERBAIJAN
Ankara
Tokat
ARMENIA
Baku
Yerevan
Gyanja
Turkmenabat
Kara Desert
TURKEY
Mt Ararat 5165 m
TURKMENISTAN
Izmir
Lake Tuz
Kayseri
Lake Van
Khvoy
Mt Sabalan 4810 m
Ashkhabad
Mary
Konya
Malatya
Tabriz
Bojnurd
Antalya
Adana
Diyarbakir
Rasht
Mersin
Gaziantep
Lake Urmia
Meshed
Rhodes
Antakya
Aleppo
Mosul
Bowkan
Babol
Sari
Sabzevar
GREECE
CYPRUS
Rakka
Erbil
Karaj
Damavand Peak 5771 m
Nicosia
Deir-ez-Zor
Kirkuk
Tehran
Kavir Desert
Limassol
Tripoli
Herat
MEDITERRANEAN SEA
Homs
Kermanshah
Qom
AFGHANISTAN
Euphrates
Tigris
ZAGROS
Beirut
SYRIA
Ba'quba
Kashan
PLATEAU OF IRAN
LEBANON
Damascus
Birjand
Haifa
Al Fallujah
Baghdad
Esfahan
Syrian Desert
IRAN
Alexandria
Tel Aviv-Jaffa
Karbala
River
Lut Desert
Port Said
ISRAEL
Amman
IRAQ
Yazd
Jerusalem
An Najaf
Amara
Masjed Soleyman
Zabol
Gaza
River
Ahvaz
MOUNTAINS
Mt Bul 3965 m
Rafsanjan
Kerman
Cairo
JORDAN
Badanah
An Nasiriya
Basra
Giza
Suez
Shiraz
Zahedan
El Faiyum
Sinai Peninsula
Sirjan
PAKISTAN
Mt Hazaran 4420 m
Kuwait
KUWAIT
Mt Sinai 2285 m
Hawiyah
Bushehr
El Minya
Tabuk
Jahrom
Nile
Hafar-al-Batin
Asyut
Persian Gulf
Bandar Abbas
Sohag
Hail
BASHAKERD MOUNTAINS
Arabian Desert
Luxor
Dammam
Manama
Buraida
EGYPT
RED SEA
BAHRAIN
Doha
Dubai
Hofuf
Idfu
River
QATAR
Gulf of Oman
Aswan
Abu Dhabi
Sohar
Muscat
Riyadh
Tropic of Cancer
Medina
Yenbo
Al Kharj
UNITED ARAB EMIRATES
Suwayq
Lake Nasser
Ibri
Mt Sham 3026m
SAUDI ARABIA
Abri
Nubian Desert
Jidda
Mecca
Arabian Peninsula
Masirah
OMAN
Port Sudan
ARABIAN SEA
Abha
Atbara
Najran
SUDAN
Atbara River
Salalah
ERITREA
Khartoum
YEMEN
INDIAN OCEAN
Asmara
Sana'a
Blue
White
Wad Medani
Hodeida
Dhamar
Mukalla
Gedaref
Nile
Mt Manar 3220 m
Kosti
Ras Dashan 4620 m
Mekele
Taizz
Nile
River
Gonder
Aden
Socotra (Yemen)
Gulf of Aden
River
Lake Tana
Bahir Dar
DJIBOUTI
Djibouti
Cape Guardafui
Dese
ETHIOPIA
Berbera
Hordio
SOUTH SUDAN
Dire Dawa
Hargeysa
SOMALIA
0 150 300 450 km
1 centimetre on the map measures 150 kilometres on the ground.
N
W E
S
A B C D E
1 2 3 4 5
30°E 40°E 50°E 60°E
10°N 20°N 30°N 40°N

Europe

Europe is made up of 45 countries. It is home to approximately 743 million people, making it the third most populated continent. Nearly 75 per cent of the population lives in cities. Europe has a rich history dating back to the ancient Greek and Roman civilisations. The idea of democracy came from the Ancient Greeks. Over 200 different languages are spoken, with German, Russian and French being the most common.

Democracy (page 64)

Eroding coastline

The sea has eroded the coastline of the island of Noss in the Shetland Islands. The sandstone rock has been shaped into caves, arches and steep cliffs. The cliffs have ledges that are used by coastal birds such as gannets, puffins and great skuas.

The changing coastline (page 24)

Europe facts	
Population:	743 000 000
Largest city:	Moscow 12 200 000 people

Winter in northern Europe

The northern half of Europe has cold winters with snow. This canal in Amsterdam has frozen. During winter, local people use skates as their transport along the canal. Boats are used for transport in the other seasons.

Seasons (page 20)

Friction (page 48)

The Eiffel Tower

Finished in 1889, the Eiffel Tower was the tallest building in the world. The 324-metre-tall tower is made of iron. It is the most visited monument in the world.

How is steel made? (page 32)

Northern Europe

Underground heat

Heat from beneath the Earth can be used to provide hot water and power. In Iceland, hot volcanic rock is close to the surface. Hot underground water (up to 150 degrees Celsius) is pumped into homes for heating.

Producing heat (page 42)

More than 400 000 people visit the Blue Lagoon geothermal pool in Iceland every year.

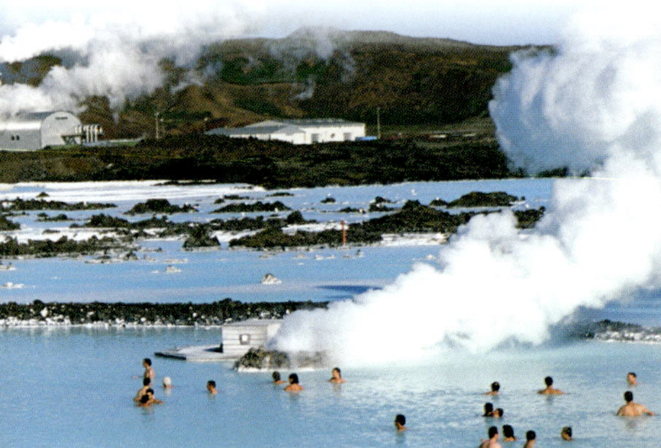

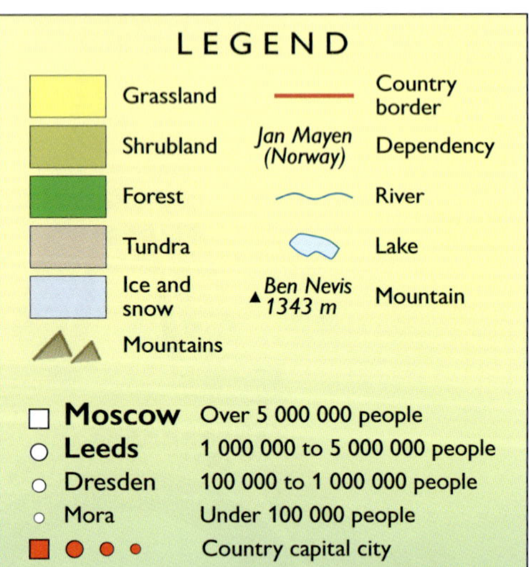

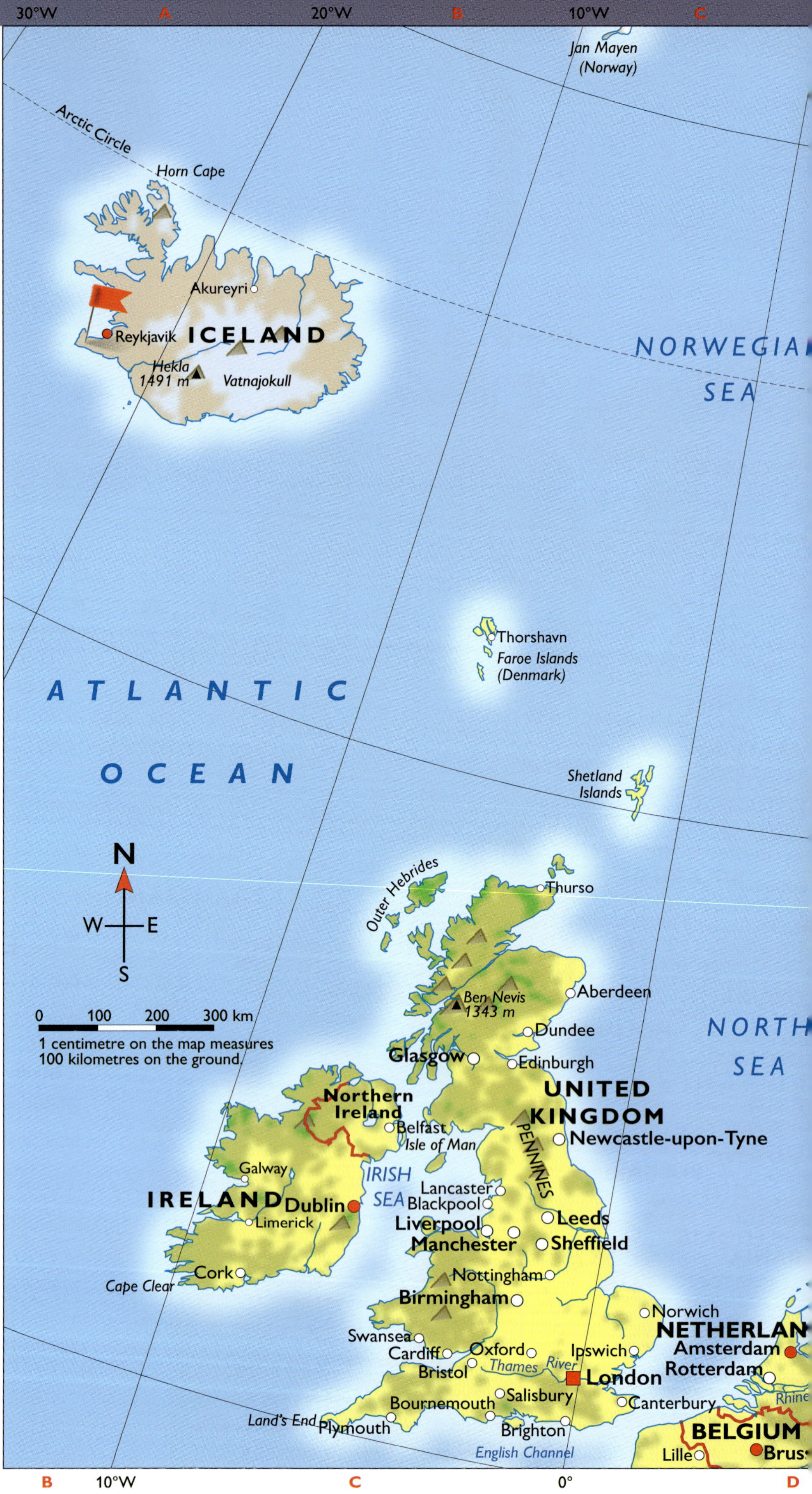

30°W
A
20°W
B
10°W
C
Jan Mayen
(Norway)
Arctic Circle
Horn Cape
Akureyri
Reykjavik
ICELAND
Hekla
1491 m
Vatnajokull
NORWEGIAN
SEA
3
60°N
Thorshavn
Faroe Islands
(Denmark)
A T L A N T I C
O C E A N
Shetland
Islands
N
W
E
S
Thurso
Outer Hebrides
2
Ben Nevis
1343 m
Aberdeen
Dundee
0
100
200
300 km
1 centimetre on the map measures
100 kilometres on the ground.
Glasgow
Edinburgh
NORTH
SEA
UNITED
KINGDOM
Northern
Ireland
Belfast
Isle of Man
Newcastle-upon-Tyne
PENNINES
Galway
IRISH
SEA
Lancaster
Blackpool
IRELAND
Dublin
Limerick
Liverpool
Leeds
Manchester
Sheffield
Cork
Cape Clear
Nottingham
Birmingham
Norwich
NETHERLANDS
Amsterdam
Rotterdam
50°N
Swansea
Cardiff
Oxford
Ipswich
Bristol
Thames
River
London
1
Salisbury
Bournemouth
Canterbury
Rhine
Land's End
Plymouth
Brighton
BELGIUM
English Channel
Lille
Brus
B
10°W
C
0°
D

LEGEND

Grassland		Country border	
Shrubland	*Jan Mayen (Norway)*	Dependency	
Forest		River	
Tundra		Lake	
Ice and snow	▲ *Ben Nevis 1343 m*	Mountain	
Mountains			

Moscow	Over 5 000 000 people
Leeds	1 000 000 to 5 000 000 people
Dresden	100 000 to 1 000 000 people
Mora	Under 100 000 people
	Country capital city

D 10°E E 20°E F 30°E G 40°E H 50°E

North Cape
BARENTS SEA
KANIN PENINSULA
Chesha Bay
Chizha
Alta
Murmansk
Lenvik
Haltiatunturi 1324 m
KOLA PENINSULA
Mezen
Kandalaksha
Kebnekaise 2123 m
Sodankyla
LAPLAND
WHITE SEA
Loukhi
Bodo
MOUNTAINS
Arkhangelsk
Jakkvik
Kem
North Dvina River
Lulea
Oulu
KJOLEN
FINLAND
Umea
Gulf of Bothnia
Stromsund
Kargopol
SCANDINAVIAN
Kuopio
Trondheim
Vaasa
Petrozavodsk
Lake Onega
60°N
Molde
PENINSULA
Harnosand
Kharovsk
NORWAY
SWEDEN
Lake Ladoga
Tampere
Glittertind 2472 m
Vologda
Lahti
Cherepovets
Mora
Helsinki
St Petersburg
Bergen
Gavle
Turku
Espoo
Gulf of Finland
Yaroslavl
Narva
Uppsala
Oslo
Vasteras
Tallinn
Luga
Velikiy Novgorod
Karlstad
Stockholm
ESTONIA
RUSSIA
Kragero
Lake Vanern
Parnu
Tartu
Tver
Egersund
Pskov
Ostashkov
Linkoping
Saaremaa
Gulf of Riga
Valga
Moscow
Lake Vattern
Skagerrak
Farosund
Goteborg
LATVIA
2
Kolomna
Boras
Jonkoping
Gotland
Riga
West Dvina River
Velikiye Luki
Serpukhov
Alborg
Kattegat
Liepaja
Vaxjo
Daugavpils
Tula
Vitebsk
DENMARK
Siauliai
Smolensk
BALTIC SEA
LITHUANIA
Copenhagen
Malmo
Kaunas
Odense
Borisov
Orel
(RUSSIA)
Vilnius
Bryansk
Kaliningrad
Minsk
Gdansk
Kiel
Oldenburg
BELARUS
Kursk
Koszalin
Rostock
Olsztyn
Baranovichi
Gomel
Salihorsk
Hamburg
Grudziadz
Groningen
Szczecin
Mozyr
Bremen
Elbe River
Vistula River
Pripet River
Chernihiv
Pinsk
Oder River
Brest
Hannover
Berlin
Poznan
Warsaw
Chornobyl'
Kharkiv
50°N
Magdeburg
Duisburg
GERMANY
Lodz
UKRAINE
Kyiv
Dortmund
Gottingen
POLAND
Lublin
Poltava
1
Dusseldorf
Leipzig
Rivne
Cologne
Dresden
Wroclaw
Kielce
Kremenchuk

10°E E 20°E F 30°E G

3

Southern Europe

Gallipoli

More than 8000 soldiers died at Gallipoli. The Australian and New Zealand soldiers landed at Anzac Cove on 25 April 1915, now remembered on ANZAC Day. They fought against the Turkish soldiers for eight months. No side could win.

Remembering the ANZACs (page 68)

Australian and New Zealand soldiers arrive at Anzac Cove, Turkey, in 1915.

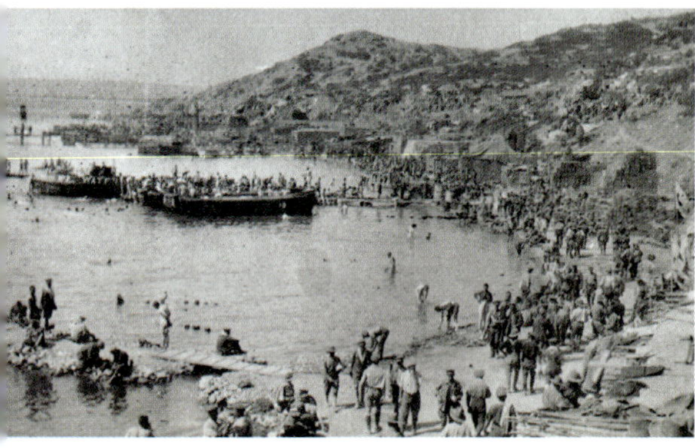

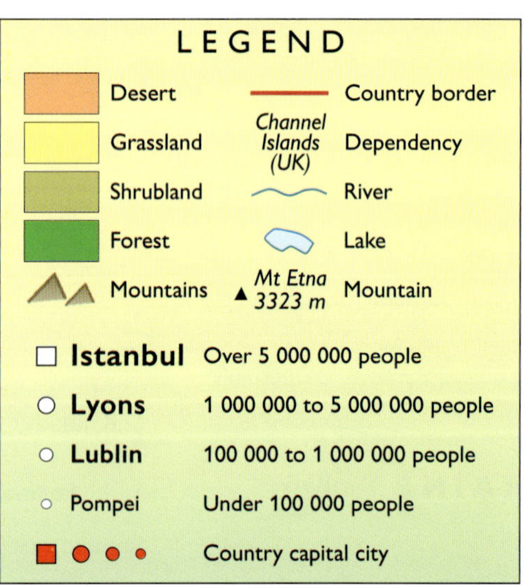

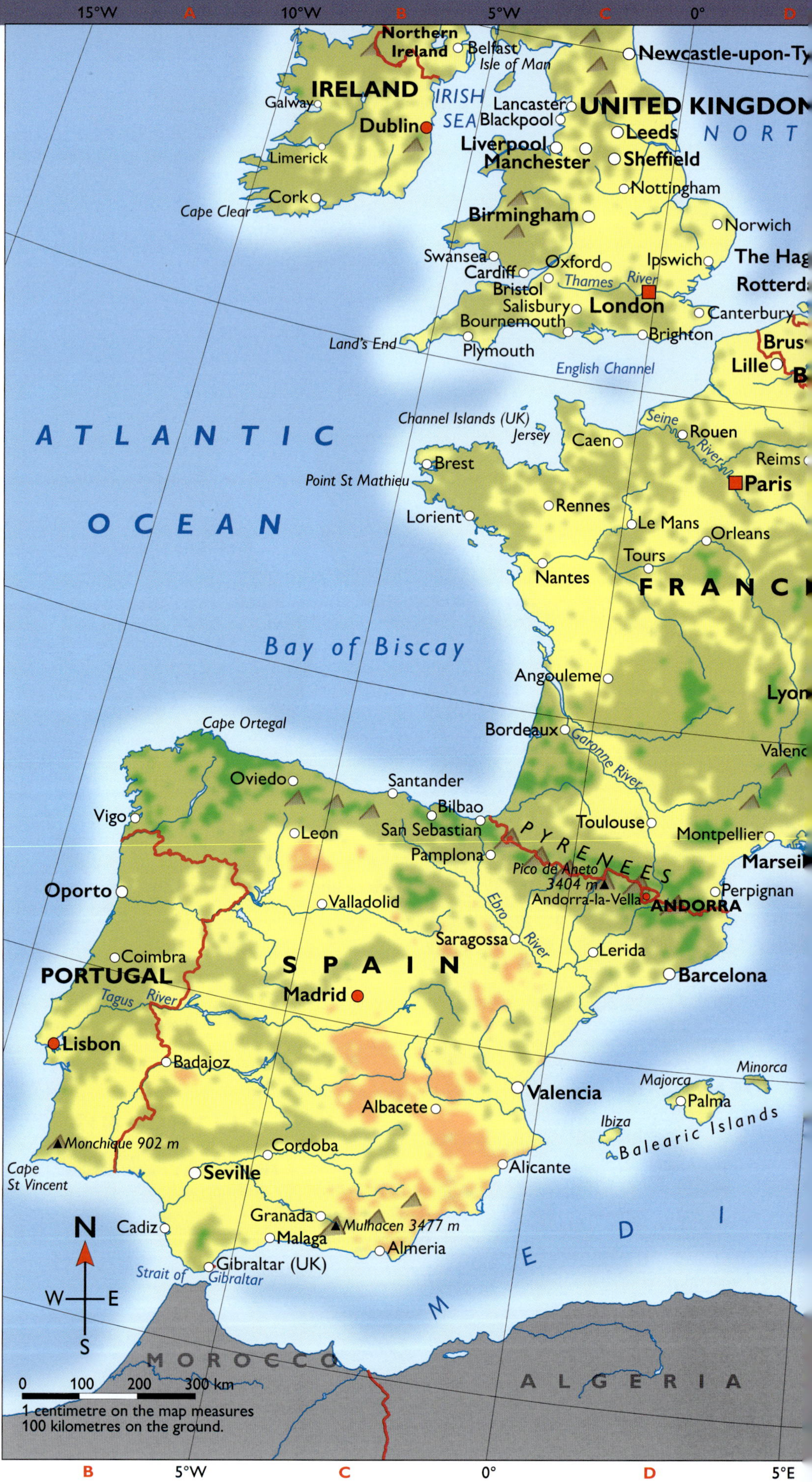

15°W
10°W
5°W
0°
5°E
50°N
45°N
40°N
35°N
A
B
C
D
1
2
3
4
5
Northern Ireland
Belfast
Isle of Man
Newcastle-upon-Ty
IRELAND
Galway
Dublin
IRISH SEA
Lancaster
Blackpool
UNITED KINGDOM
Leeds
N O R T
Liverpool
Manchester
Sheffield
Limerick
Nottingham
Cork
Cape Clear
Birmingham
Norwich
Swansea
Cardiff
Oxford
Ipswich
The Hag
Rotterd
Bristol
Thames River
Salisbury
London
Canterbury
Bournemouth
Brighton
Land's End
Plymouth
English Channel
Brus
Lille
A T L A N T I C
O C E A N
Channel Islands (UK)
Jersey
Caen
Rouen
Seine River
Reims
Brest
Point St Mathieu
Paris
Rennes
Lorient
Le Mans
Orleans
Tours
Nantes
F R A N C
Bay of Biscay
Angouleme
Lyon
Bordeaux
Garonne River
Valenc
Cape Ortegal
Oviedo
Santander
Bilbao
Vigo
Leon
San Sebastian
Toulouse
Montpellier
Pamplona
P Y R E N E E S
Marsei
Pico de Aneto 3404 m
Perpignan
Oporto
Andorra-la-Vella
ANDORRA
Valladolid
Ebro River
Saragossa
Lerida
Coimbra
PORTUGAL
S P A I N
Barcelona
Madrid
Tagus River
Lisbon
Badajoz
Minorca
Majorca
Palma
Valencia
Albacete
Ibiza
Balearic Islands
Monchique 902 m
Cordoba
Cape St Vincent
Seville
Alicante
Granada
Mulhacen 3477 m
Cadiz
Malaga
Almeria
Gibraltar (UK)
Strait of Gibraltar
N
W
E
S
M E D I
M O R O C C O
A L G E R I A
0 100 200 300 km
1 centimetre on the map measures 100 kilometres on the ground.

LEGEND

Desert		Country border	
Grassland	*Channel Islands (UK)*	Dependency	
Shrubland		River	
Forest		Lake	
Mountains	▲ *Mt Etna 3323 m*	Mountain	

Istanbul	Over 5 000 000 people
Lyons	1 000 000 to 5 000 000 people
Lublin	100 000 to 1 000 000 people
Pompei	Under 100 000 people
	Country capital city

DENMARK
SWEDEN
BALTIC SEA
LITHUANIA
(RUSSIA)
BELARUS
RUSSIA
POLAND
GERMANY
NETHERLANDS
LUXEMBOURG
CZECH REPUBLIC
SLOVAKIA
UKRAINE
MOLDOVA
AUSTRIA
HUNGARY
ROMANIA
LIECHTENSTEIN
SWITZERLAND
SLOVENIA
CROATIA
BOSNIA and HERZEGOVINA
SERBIA
BULGARIA
MONTENEGRO
KOSOVO
NORTH MACEDONIA
ALBANIA
GREECE
TURKEY
ITALY
SAN MARINO
MONACO
VATICAN CITY
MALTA
TUNISIA
ADRIATIC SEA
TYRRHENIAN SEA
IONIAN SEA
AEGEAN SEA
SEA OF CRETE
BLACK SEA
TRANSYLVANIAN ALPS
Copenhagen
Malmo
Kaunas
Vilnius
Borisov
Minsk
Bryansk
Kaliningrad
Gdansk
Koszalin
Olsztyn
Baranovichi
Salihorsk
Gomel
Mozyr
Chernihiv
Kiel
Oldenburg
Rostock
Hamburg
Bremen
Szczecin
Grudziadz
Pinsk
Brest
Chornobyl'
Warsaw
Poznan
Lodz
Lublin
Kyiv
Amsterdam
Hannover
Berlin
Magdeburg
Ruhr
Dortmund
Essen
Dusseldorf
Cologne
Gottingen
Leipzig
Dresden
Wroclaw
Kielce
Rivne
Kremenchuk
Koblenz
Frankfurt
Katowice
Krakow
L'viv
Ternopil'
Vinnytsya
Prague
Ostrava
Luxembourg
Mannheim
Nuremberg
Gerlachovka 2655 m
Uzhhorod
Nancy
Stuttgart
Bratislava
Vienna
Baia-Mare
Iasi
Kishinev
Odesa
Freiburg
Munich
Salzburg
Gyor
Budapest
Debrecen
Bacau
Basel
Zurich
Vaduz
Innsbruck
Cluj
Bern
Klagenfurt
Lake Balaton
Arad
Braila
Lake Geneva
Mt Blanc 4807 m
Trent
Ljubljana
Pecs
Zagreb
Resita
Bucharest
Constanta
Milan
Venice
Rijeka
Novi Sad
Belgrade
Grenoble
Turin
Banja Luka
Tuzla
Craiova
Ruse
Varna
Genoa
La Spezia
Bologna
Sarajevo
Nis
Burgas
Nice
Monaco
Pisa
Florence
San Marino
Split
Sofia
Toulon
Perugia
Pristina
Plovdiv
Istanbul
Corlu
Tekirdag
Podgorica
Skopje
Bastia
Corsica
Pescara
Alexandroupolis
Kavalla
Rome
Tirane
Durres
Latina
Bari
Salonika
Brindisi
Naples
Pompeii
Vesuvius 1281 m
Salerno
Mt Olympus 2917 m
Sassari
Sardinia
Gulf of Taranto
Corfu
Chios
Cosenza
Catanzaro
Khalkis
Agrinion
Athens
Cagliari
Cape Teulada
Ionian Islands
Patras
Corinth
Ikaria
Messina
Palermo
Naxos
Rhodes
Mt Etna 3323 m
Catania
Cape Spartivento
Sparta
Sicily
Siracusa
Cape Tainaron
Iraklion
Cape Spatha
Canea
Crete
Valletta
Danube
Elbe River
Oder River
Vistula River
Rhine River
Tisza River
Tiber River
Dnister River
Prut River
Dnipro River
Pripet River
APENNINES
ALPS

Russia and neighbours

Woolly mammoth

Remains of the extinct woolly mammoth have been found across northern Siberia. In 2012, a 30 000-year-old carcass was found on the Taimyr Peninsula. It still had some of its organs, and pieces of flesh and fur.

Extinct animals (page 36)

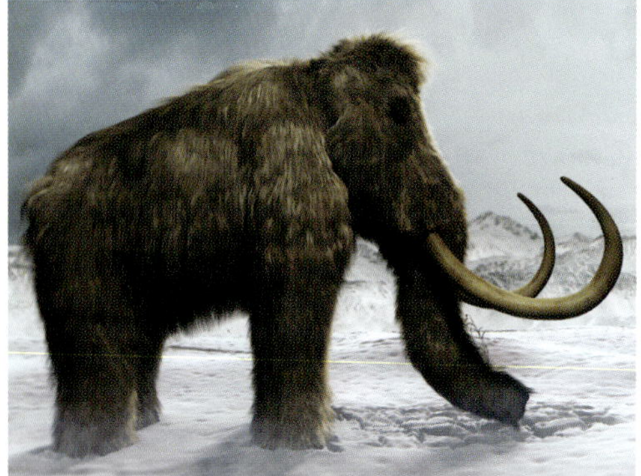

Woolly mammoths were about the same size as African elephants.

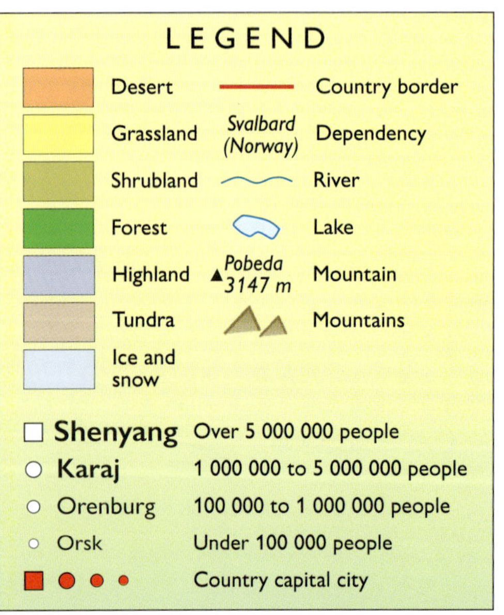

LEGEND

Desert	Country border
Grassland	Svalbard (Norway) Dependency
Shrubland	River
Forest	Lake
Highland	Pobeda 3147 m Mountain
Tundra	Mountains
Ice and snow	

Shenyang	Over 5 000 000 people
Karaj	1 000 000 to 5 000 000 people
Orenburg	100 000 to 1 000 000 people
Orsk	Under 100 000 people
	Country capital city

50°N
60°N
70°N
10°E
20°E
30°E
40°E
40°N
50°E
30°N
60°E
70°E
80°E
NETHERLANDS
DENMARK
Oslo
NORWAY
NORWEGIAN SEA
Hannover
Hamburg
GERMANY
Copenhagen
SWEDEN
North Cape
Berlin
Stockholm
Prague
Gotland
CZECH REPUBLIC
BALTIC SEA
Turku
Oulu
FINLAND
BARENTS SEA
POLAND
Gdansk
Helsinki
Murmansk
Lodz
Warsaw
(RUSSIA)
Riga
Tallinn
ESTONIA
LITHUANIA
Tartu
Lublin
Vilnius
LATVIA
Lake Ladoga
WHITE SEA
Pskov
St Petersburg
Nova
Kolguyev
L'viv
Velikiy Novgorod
Minsk
BELARUS
Arkhangelsk
Kara Strait
Vaygach Island
Smolensk
TIMAN RIDGE
Gomel
Vologda
Kyiv
Moscow
Yaroslavl
MOLDOVA
Vorkuta
Tula
Ivanovo
URAL MOUNTAINS
UKRAINE
Ryazan
Syktyvkar
Mt Narodnaya 1894 m
Kharkiv
Nizhny Novgorod
Voronezh
R
Donets'k
Glazov
Kazan
Mt Konzhakovskiy 1569 m
Sea of Azov
Ul'yanovsk
Perm
BLACK SEA
Rostov
Saratov
Naberezhnyye Chelny
Krasnodar
Samara
Ob
Volgograd
Sochi
Ufa
Yekaterinburg
Surg
Stavropol
Volga River
Uralsk
Chelyabinsk
Tyumen
Mt Elbrus 5642 m
Irtysh River
WEST SIBERIAN PLAIN
Orenburg
Magnitogorsk
GEORGIA
Astrakhan
Grozny
Aktyubinsk
Orsk
Kustanai
Tbilisi
Makhachkala
KIRGHIZ STEPPE
Omsk
Yerevan
ARMENIA
KAZAKHSTAN
Aktau
Chelkar
AZERBAIJAN
CASPIAN SEA
UST URT PLATEAU
ARAL SEA
Astana
Baku
Barna
Karaganda
Rasht
UZBEKISTAN
Amu
Syr
Semipalatinsk
Kzyl-Orda
Dashoguz
TURKMENISTAN
Kyzyl Desert
River
Tehran
Kara Desert
Lake Balkhash
Ashkhabad
IRAN
Chimkent
Taraz
Mary
Tashkent
Bishkek
Alma-Ata
Meshed
Samarkand
Namangan
Yining
Yazd
Dushanbe
KYRGYZSTAN
TAJIKISTAN
Urumc
Birjand
Herat
Communism Peak 7495 m
Kashi
Lut Desert
Baghlan
Muztagh Ata 7546 m
HINDU KUSH
Yarkant River
Zabol
AFGHANISTAN
TARIM BASIN
Margo Desert
Kabul
Zahedan
Kandahar
Peshawar
PAKISTAN
Godwin Austen (K2) 8611 m
Hotan
1
2
3
4
5
6
G
H
I

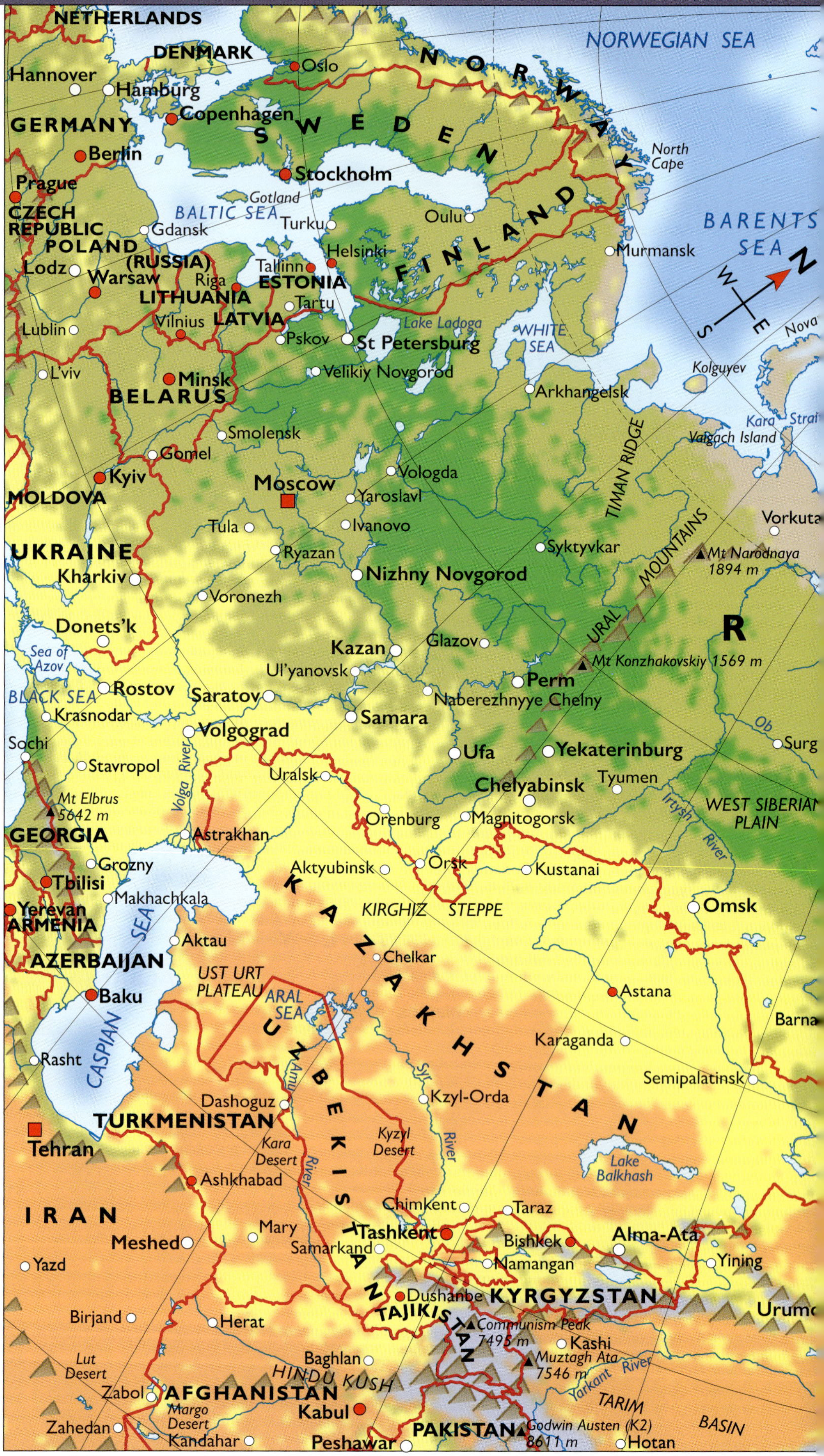

ARCTIC OCEAN
BEAUFORT SEA
0 200 400 600 km
1 centimetre on the map measures 225 kilometres on the ground.
Alaska (USA)
Nome
Bering Strait
CHUKCHI SEA
Wrangel Island
Chukot Peninsula
Arctic Circle
CHUKOT RANGE
Cape Shelagski
Ayon Island
Cape Navarin
EAST SIBERIAN SEA
New Siberian Islands
Lyakhov Island
Franz Josef Land
Cape Arkticheskiy
Severnaya Zemlya
Bolshevik Island
KARA SEA
LAPTEV SEA
Delta of the Lena
Taimyr Peninsula
Lake Taimyr
Dikson
Nordvik
Tiksi
Chokurdakh
Cherskiy
Markovo
Mt Ledyanaya 2562 m
BERING SEA
NORTH SIBERIAN LOWLANDS
Norilsk
PUTORANA MOUNTAINS
RUSSIA
CHERSKI RANGE
Pobeda 3147 m
KOLYMA RANGE
Gulf of Shelikhov
SREDINNY RANGE
Karagin Island
Klyuchevskaya Sopka 4749 m
Kamchatka Peninsula
Magadan
Mus-Khaya 2959 m
CENTRAL SIBERIAN PLATEAU
Tura
Sangar
Yakutsk
Petropavlovsk-Kamchatski
SEA OF OKHOTSK
Bakhta
Mirnyy
Siberia
DZUGDZHUR RANGE
Shantar Islands
Cape Yelizavety
Sakhalin
Kuril Islands
Lesosibirsk
Tomsk
Novosibirsk
Krasnoyarsk
Novokuznetsk
Grandioznyy 2922 m
PATOM HIGHLANDS
Lake Baikal
BUREISKIY RANGE
Yuzhno-Sakhalinsk
Munku Sardyk 3491 m
Irkutsk
Chita
GREATER HINGGAN MOUNTAINS
Birobidzhan
Khabarovsk
SIKHOTE MOUNTAINS
Mt Belukha 4503 m
Asahikawa
Kushiro
Sapporo
Hokkaido
Yichun
Qiqihar
Harbin
Mudanjiang
Jixi
Asaralta 2776 m
Ulan Bator
ALTAI MOUNTAINS
KHANGAI MOUNTAINS
MONGOLIA
Vladivostok
Hachinohe
Changchun
Jilin
Chongjin
JAPAN
Sendai
SEA OF JAPAN
Nagaoka
Honshu
Xilin Hot
Shenyang
Anshan
Gobi Desert
NORTH KOREA
Tokyo
Yokohama
Shizuoka
Pyongyang
SOUTH KOREA
Nagoya
TURFAN DEPRESSION
CHINA
Baotou
Hohhot
Beijing
Lena River
Amga River
Amur River
Zeya River
Anadyr River
Indigirka River
Taz River
100°E 110°E 120°E 130°E 140°E
80°N 70°N 60°N 50°N 40°N
170°W 180° 170°E 160°E 150°E

Africa

Africa is the second-largest continent on Earth. It is made up of 54 different countries. Africa is home to over 1.2 billion people. There are between 1000 and 2500 different languages spoken across Africa.

The largest desert in the world, the Sahara, covers most of northern Africa. The world's longest river, the Nile River, is located in Africa.

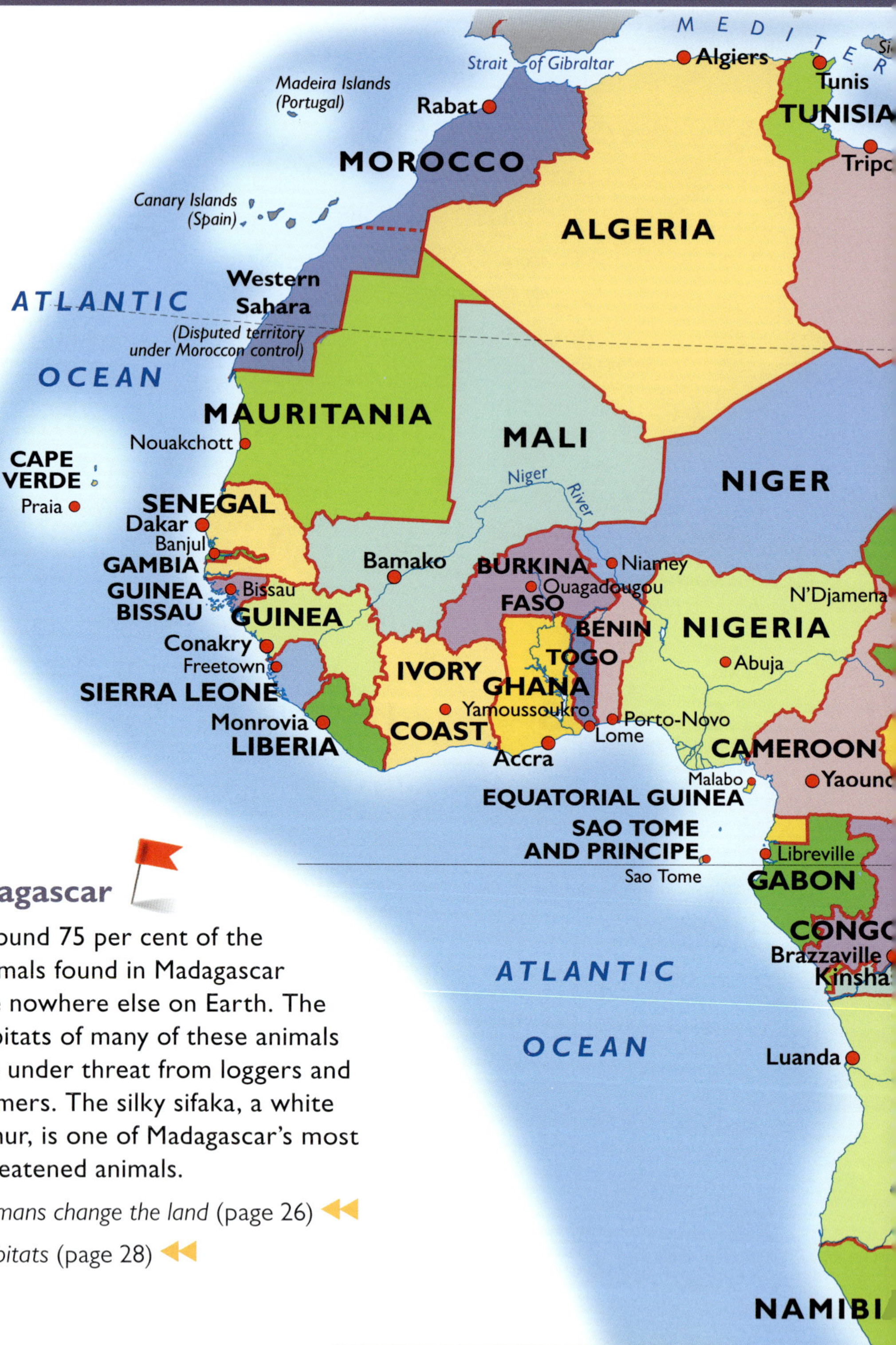

Madagascar

◁ Around 75 per cent of the animals found in Madagascar live nowhere else on Earth. The habitats of many of these animals are under threat from loggers and farmers. The silky sifaka, a white lemur, is one of Madagascar's most threatened animals.

Humans change the land (page 26) ◀◀

Habitats (page 28) ◀◀

Africa facts	
Population:	1 200 000 000
Largest city:	Cairo 19 130 000 people

Gold mining

▲ South Africa has some of the world's largest gold mines. This miner is drilling for gold deep beneath the Earth's surface.

Gold in Australia (page 31) ◀◀

World's tallest animal

▼ The world's tallest animal is the giraffe. A baby giraffe is two metres tall at birth. The giraffe is found with the world's largest land animal (the elephant) in eastern and southern Africa.

Growth (page 34) ◀◀

Life cycles (page 38) ◀◀

Northern Africa

Chimpanzees

Chimpanzees are the closest relatives to humans. They live in the forest around the Congo River.

Life cycles (page 38)
Habitats (page 28)

Chimpanzee mother and baby, Democratic Republic of Congo

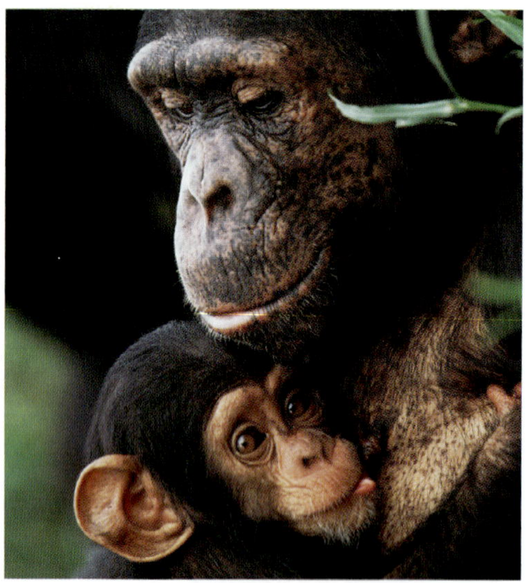

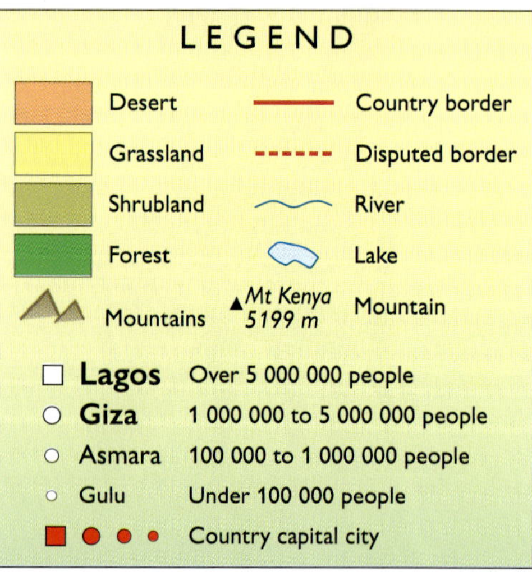

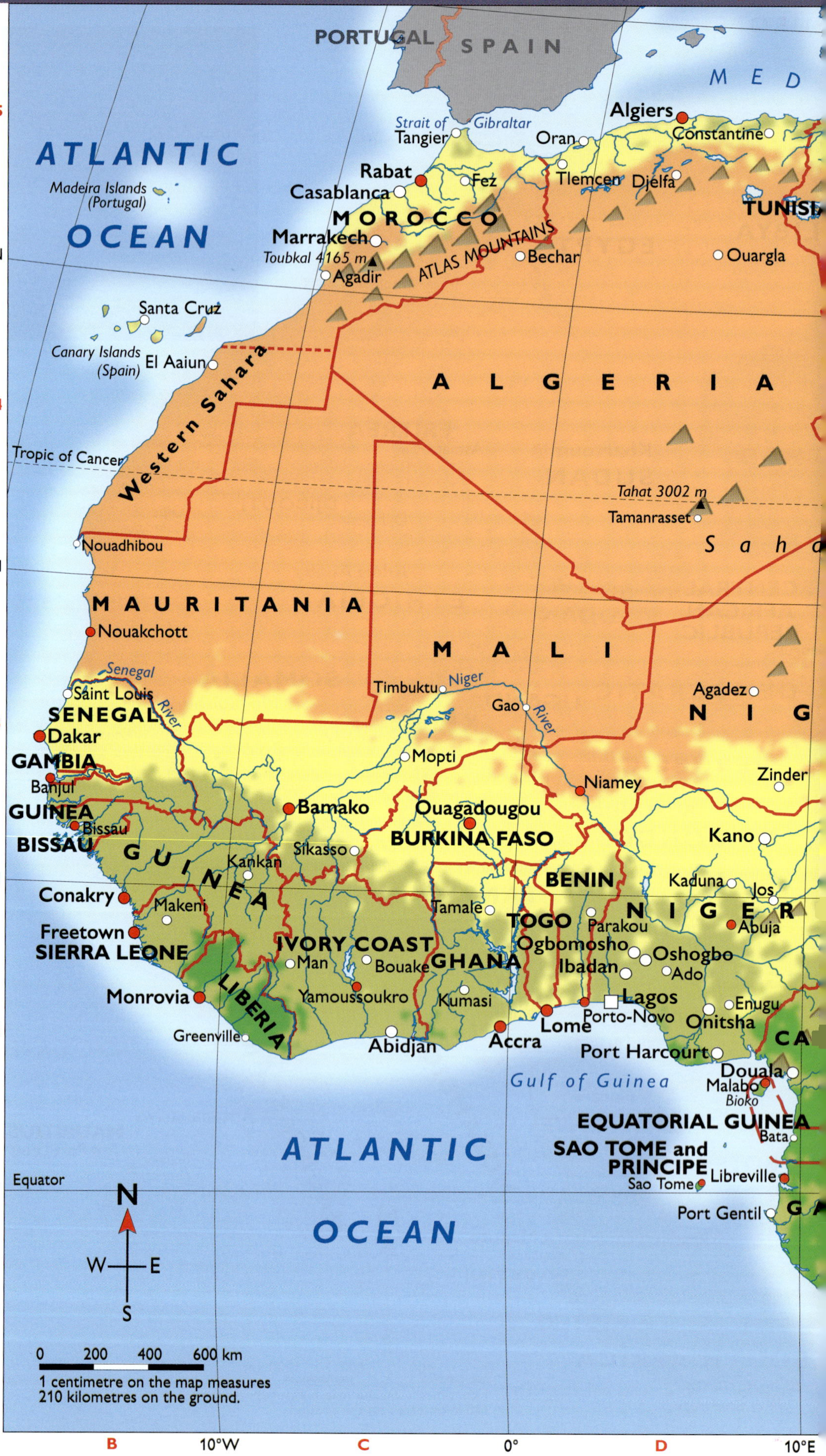

LEGEND

- Desert
- Grassland
- Shrubland
- Forest
- Mountains
- Country border
- Disputed border
- River
- Lake
- *Mt Kenya 5199 m* Mountain

Lagos	Over 5 000 000 people
Giza	1 000 000 to 5 000 000 people
Asmara	100 000 to 1 000 000 people
Gulu	Under 100 000 people
	Country capital city

°E E 20°E F 30°E G 40°E H 50°E I 60°E

Sardinia
ITALY
EUROPE
Sicily
GREECE
Tunis
Sousse
MALTA
Gabes
MEDITERRANEAN SEA
Crete
TURKEY
CYPRUS
LEBANON
SYRIA
ASIA
IRAN
IRAQ
ISRAEL
JORDAN
Tripoli
Al Khums
Al Bayda
Benghazi
Tubruq (Tobruk)
Nile Delta
Alexandria
Port Said
Cairo
Giza
Suez
El Faiyum
QATTARA DEPRESSION
Mt Sinai 2285 m
El Minya
Asyut
LIBYA
Libyan Desert
EGYPT
Qena
Luxor
Murzuq
Nile
Aswan
Persian Gulf
BAHRAIN
QATAR
UNITED ARAB EMIRATES
SAUDI ARABIA
Sahara Desert
TIBESTI MOUNTAINS
Emi Koussi 3415 m
Faya-Largeau
RED SEA
Port Sudan
Merowe
River
NIGER
CHAD
Darfur
Khartoum
ERITREA
Asmara
YEMEN
SUDAN
Blue
Wad Medani
Abeche
El Obeid
Kosti
Assab
Gulf of Aden
Socotra (Yemen)
N'Djamena
Nyala
Gonder
DJIBOUTI
Djibouti
Cape Guardafui
Maiduguri
Nile
River
Bahir Dar
Dese
VALLEY
Berbera
Kaele
White
Dire Dawa
Hargeysa
Bender Beila
Addis Ababa
Harer
Moundou
ETHIOPIA
Sudd
Wau
Jima
RIFT
Ngaoundere
SOUTH SUDAN
Ogaden Desert
CENTRAL AFRICAN REPUBLIC
Batu 4307 m
Bambari
CAMEROON
GREAT
SOMALIA
Ubangi
Juba
Bangui
River
Yaounde
Lake Turkana
Gemena
Gulu
Isiro
River
Mogadishu
DEMOCRATIC
Bunia
UGANDA
KENYA
Margherita 5109 m
Kampala
Mbandaka
Kisangani
Eldoret
Mt Kenya 5199 m
GABON
Lake Edward
Lake Victoria
Kisumu
Kismayu
Libreville
REPUBLIC
Tana River
Goma
Nairobi
CONGO
Congo
Lake Mai-Ndombe
Lualaba
Kigali
RWANDA
Kindu
Bukavu
Mt Kilimanjaro 5895 m
Bandundu
BURUNDI
Arusha
INDIAN
Brazzaville
OF CONGO
Bujumbura
Shinyanga
Mombasa
Pointe Noire
Kasai
Ilebo
Kinshasa
River
River
Kigoma
Kikwit
Tabora
Cabinda (ANGOLA)
Boma
Kananga
Kalemie
TANZANIA
Mbuji-Mayi
Zanzibar
OCEAN
Tshikapa
Dodoma
Lake Tanganyika
Dar-es-Salaam
Mwene-Ditu
ANGOLA
Iringa

5 30°N 4 20°N 3 10°N 2 0° 1

E 20°E F 30°E G 40°E H 50°E I

Southern Africa

Dead as a dodo

The dodo is an extinct bird that lived on the island of Mauritius. The dodo was a flightless bird and easy to catch. When people first landed on Mauritius in 1598, the dodo was killed for meat. Settlers also brought dogs, cats and pigs. Many of these animals escaped and ate the dodos and their eggs. After less than 100 years of human settlement, the dodo became extinct in 1681.

Extinct animals (page 36) ◀◀ *Threatened species* (page 55) ◀◀

This painting by Roelant Savery (1626) helps us to understand what a dodo looked like.

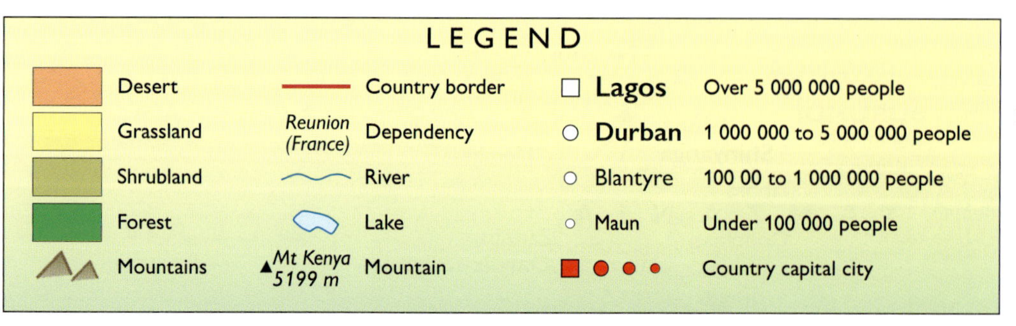

SUDAN
Nyala
CHAD
CENTRAL AFRICAN REPUBLIC
Bambari
SOUTH SUDAN
Wau
Sudd
Juba
White Nile River
ETHIOPIA
Bahir Dar
Dese
Addis Ababa
Dire Dawa
Harer
Jima
Batu 4307 m
GREAT RIFT VALLEY
DJIBOUTI
Djibouti
Gulf of Aden
Socotra (Yemen)
Cape Guardafui
Berbera
Hargeysa
Bender Beila
Ogaden Desert
SOMALIA
Mogadishu
Kismayu
Lake Turkana
KENYA
Gulu
UGANDA
Kampala
Bunia
Margherita 5109 m
Eldoret
Mt Kenya 5199 m
Lake Victoria
Kisumu
Nairobi
Tana River
Mt Kilimanjaro 5895 m
Mombasa
Arusha
Shinyanga
DEMOCRATIC REPUBLIC OF CONGO
Gemena
Isiro
Congo River
Ubangi River
Kisangani
Mbandaka
Lake Mai-Ndombe
Goma
Kigali
RWANDA
Bukavu
Kindu
BURUNDI
Bujumbura
Bandundu
Ilebo
Kasai River
Kikwit
Kananga
Mbuji-Mayi
Mwene-Ditu
Tshikapa
Lualaba River
Kigoma
Tabora
Kalemie
Lake Tanganyika
TANZANIA
Dodoma
Zanzibar
Dar-es-Salaam
Iringa
Mbeya
INDIAN OCEAN
Victoria
SEYCHELLES
Lake Mweru
Kolwezi
Likasi
Lubumbashi
Livingstonia
Songea
Mtwara
COMOROS
Moroni
Cape Ambre
Mayotte (France)
Maromokotro 2876 m
ANGOLA
Luena
PLATEAU
Menongue
Kitwe
ZAMBIA
Chipata
Lake Malawi
MALAWI
Lilongwe
Kabwe
Lusaka
Nampula
Mahajanga
Tete
Blantyre
Zambezi River
Mocuba
Lake Kariba
Livingstone
Victoria Falls
Harare
ZIMBABWE
Gweru
Bulawayo
Chimoio
Beira
Mozambique Channel
MOZAMBIQUE
MADAGASCAR
Toamasina
Antananarivo
MAURITIUS
Port Louis
St-Denis
Reunion (France)
Fianarantsoa
Okavango Delta
Maun
Makgadikgadi Pan
NAMIBIA
Windhoek
BOTSWANA
Kalahari Desert
Europa (France)
Toliara
Maxixe
Gaborone
Pretoria
Xai-Xai
Mbombela
Johannesburg
Mbabane
Maputo
ESWATINI
Klerksdorp
Vereeniging
Welkom
Newcastle
Kimberley
Orange River
Bloemfontein
Maseru
Pietermaritzburg
LESOTHO
Durban
Cape Sainte-Marie
SOUTH AFRICA
East London
Port Elizabeth
Cape Town
Cape of Good Hope
Cape Agulhas
20°E
30°E
40°E
50°E
60°E
10°N
0°
10°S
20°S
30°S
C
D
E
F
1
2
3
4
5
6

North America

North America is made up of 22 countries—Canada and the United States of America (USA) are the second- and third-largest countries in the world. Its people most commonly speak English, Spanish and French. Christianity is North America's most popular religion.

Colorado River

The Colorado River in south-western United States of America is over 2330 kilometres long. It begins in the Rocky Mountains and flows into the Gulf of Mexico. The Grand Canyon, the world's largest gorge, has been carved out by the Colorado River over millions of years.

Rivers change the land (page 22)

BERING SEA

Aleutian Islands

Alaska (USA)

Yukon River

LEGEND

(red line)	Country border
HAITI	Country name
Puerto Rico (USA)	Dependency
Country capital city	
■ **Mexico City**	Over 5 000 000 people
● **Ottawa**	1 000 000 to 5 000 000 people
● Nassau	100 000 to 1 000 000 people
• Belmopan	Under 100 000 people

PACIFIC OCEAN

Hawaii (USA)

Colorado

L'Anse aux Meadows

Located on the northern tip of Newfoundland, L'Anse aux Meadows contains the earliest evidence of European settlement in North America. The remains of an 11th-century Viking settlement have been found.

Leif Eriksson (page 56)

North America facts	
Population:	567 000 000
Largest city:	Mexico City 21 200 000 people

ARCTIC OCEAN
BEAUFORT SEA
Kalaallit Nunaat (Greenland) (Denmark)
Arctic Circle
LABRADOR SEA
Hudson Bay
CANADA
St Pierre and Miquelon (France)
Ottawa
Lake Superior
Missouri River
UNITED STATES OF AMERICA
Washington DC
Mississippi River
Rio Grande
Bermuda (UK)
ATLANTIC OCEAN
BAHAMAS
Nassau
Tropic of Cancer
Gulf of Mexico
MEXICO
Mexico City
Havana
CUBA
DOMINICAN REPUBLIC
Puerto Rico (USA)
ANTIGUA and BARBUDA
HAITI
Port-au-Prince
Santo Domingo
ST KITTS and NEVIS
Guadeloupe (France)
DOMINICA
Martinique (France)
BARBADOS
JAMAICA
Kingston
ST LUCIA
ST VINCENT and the GRENADINES
GRENADA
Belmopan
BELIZE
GUATEMALA
Guatemala City
HONDURAS
Tegucigalpa
CARIBBEAN SEA
San Salvador
EL SALVADOR
NICARAGUA
Managua
San Jose
COSTA RICA
Panama City
PANAMA
0 400 800 1200 km
1 centimetre on the map measures 370 kilometres on the ground.
N W E S

Bay of Fundy

The biggest change in water depth between low and high tides occurs in the Bay of Fundy in Canada. The narrow bay allows changes in water depth of up to 17 metres. The gravitational pull between the Moon and the Earth causes high and low tides. These boats rise up and down with the change of tide in the Bay of Fundy.

The Moon (page 17)
Gravity (page 47)

Death Valley

Death Valley is the hottest place in the world. Temperatures over 49 degrees Celsius are common during the summer months of June, July and August. The record temperature for Death Valley is 56.7 degrees Celsius. This sidewinder snake is moving through the desert sand in Death Valley.

World's hottest place (page 40)
Reptiles and conduction (page 45)

Canada, Alaska and Greenland

North Magnetic Pole

Compass needles point to the North Magnetic Pole. It is near, but not in the same place as the geographic North Pole. Magnetic poles move. At the moment, the Magnetic North Pole is moving at 60 kilometres per year.

Magnets (page 50)

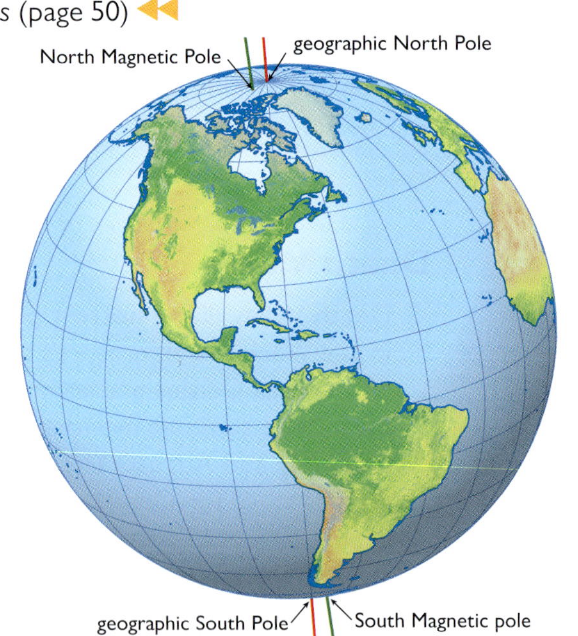

Location of Earth's poles

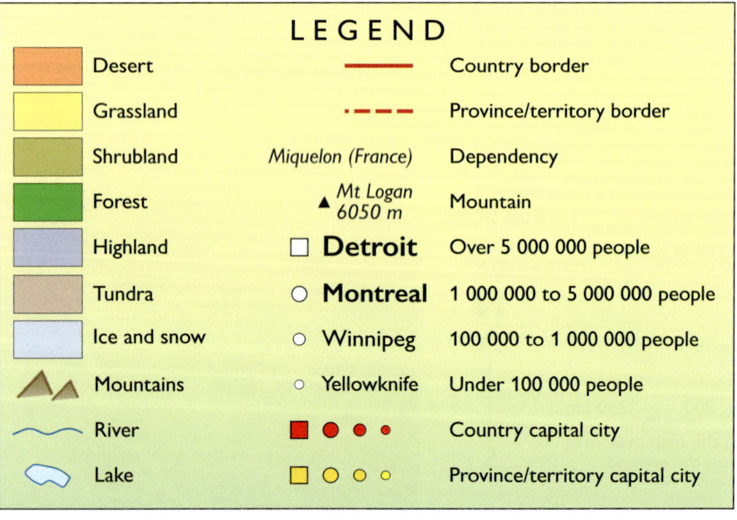

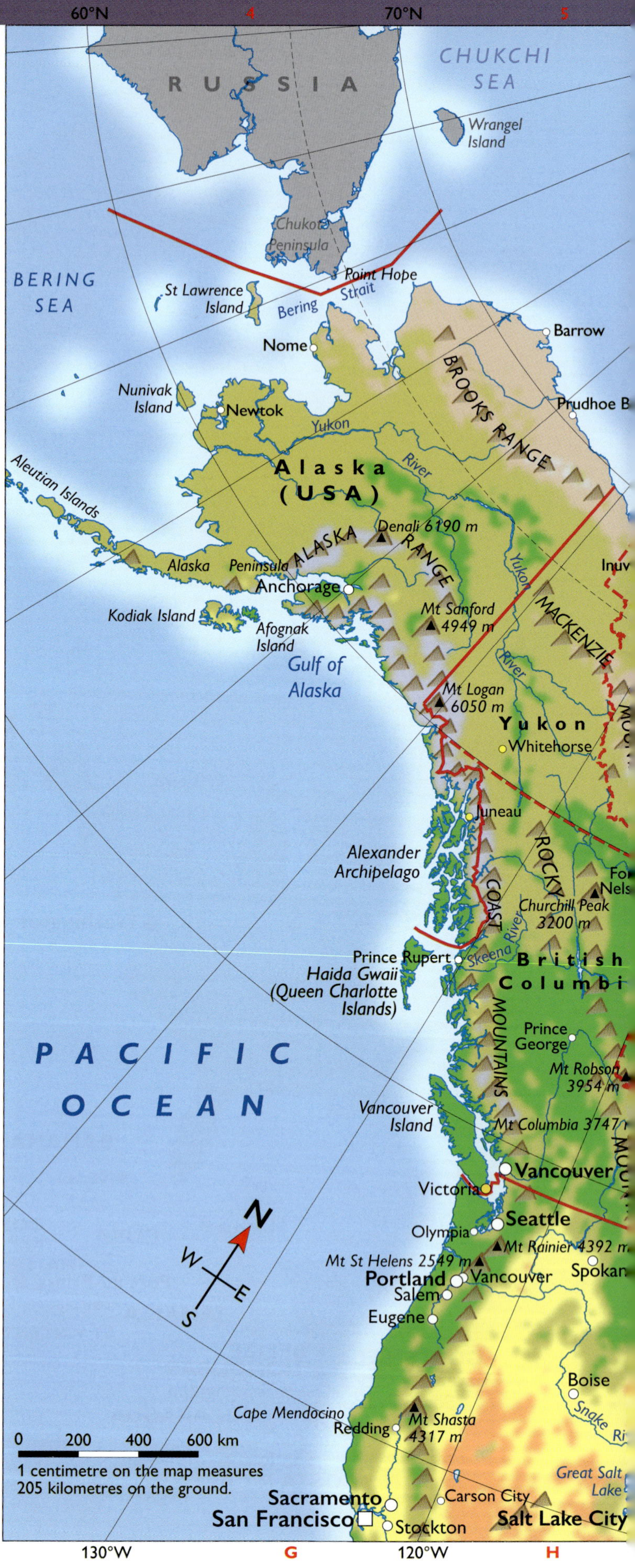

ARCTIC OCEAN
North Magnetic Pole
Queen Elizabeth Islands
Sverdrup Islands
Axel Heiberg
Ellesmere Island
Mackenzie King Island
Prince Patrick Island
Cape Prince Alfred
BEAUFORT SEA
Melville Island
Bathurst Island
Parry Islands
Cornwallis Island
Devon Island
Banks Island
Prince of Wales Island
Somerset Island
Brodeur Peninsula
Bylot Island
Baffin Bay
Qaanaaq
Kalaallit Nunaat (Greenland) (Denmark)
Jan Mayen Island (Norway)
Eskimonæs
Faroe Islands (Denmark)
ICELAND
Hekla 1491 m
Reykjavik
Horn Cape
Qeqertarsuaq
Arctic Circle
King Frederick VI Coast
Davis Strait
Nuuk
Paamiut
Nunap Isua
Victoria Island
Boothia Peninsula
Baffin Island
Great Bear Lake
Cambridge Bay
King William Island
Pelly Bay
Prince Charles Island
Foxe Basin
LABRADOR SEA
Nunavut
Coppermine River
Northwest Territories
Barren Grounds
Iqaluit
Southampton Island
Resolution Island
Cape Chidley
ATLANTIC OCEAN
Great Slave Lake
Yellowknife
Kazan River
Coats Island
Mansel Island
Ungava Peninsula
George River
Nain
Newfoundland
Labrador
Smallwood Reservoir
Alberta
Lake Athabasca
Churchill
Hudson Bay
Inukjuak
Reindeer Lake
Belcher Islands
Gros Morne 806 m
St John's
Newfoundland
Edmonton
Saskatchewan
Manitoba
Saskatchewan River
Akimiski Island
Quebec
Anticosti Island
Gulf of St Lawrence
St Pierre (France)
Miquelon (France)
Kicking Horse Pass
Calgary
Saskatoon
CANADA
Lake Winnipeg
Fort Albany
Prince Edward Island
Sydney
Cape Breton Island
Charlottetown
Regina
Ontario
Alma
St Lawrence River
New Brunswick
Nova Scotia
Winnipeg
Cochrane
Quebec
Fredericton
St John
Halifax
Thunder Bay
Trois Rivieres
Bay of Fundy
Helena
Missouri River
Lake Superior
Sudbury
Montreal
Ottawa
Augusta
Yellowstone River
Bismarck
Fargo
Duluth
Montpelier
Concord
Manchester
Billings
Lake Huron
Lake Michigan
Lake Ontario
Toronto
Boston
St Cloud
Minneapolis
St Paul
Albany
Rochester
Providence
Pierre
Grand Rapids
London
Buffalo
Hartford
Long Island
UNITED STATES OF AMERICA
Milwaukee
Madison
Lake Erie
New York
Detroit
Chicago
Cleveland
Pittsburgh
Philadelphia
Cheyenne
N E S W

80°N 80°N 70°N 0° 10°W 20°W 30°W 40°W 50°N 50°W 40°N 60°W
180° 160°W 140°W 120°W 100°W 80°W 60°W 40°W 20°W
80°W 100°W 90°W 80°W 70°W
A B C D E F G H I J K L M N O P Q R S
1 2 3 4 5 6

United States of America

Disneyland rides

The car on the log flume ride at Disneyland suddenly drops and speeds towards the ground under the influence of gravity.

Gravity (page 47)

The log flume ride at Disneyland

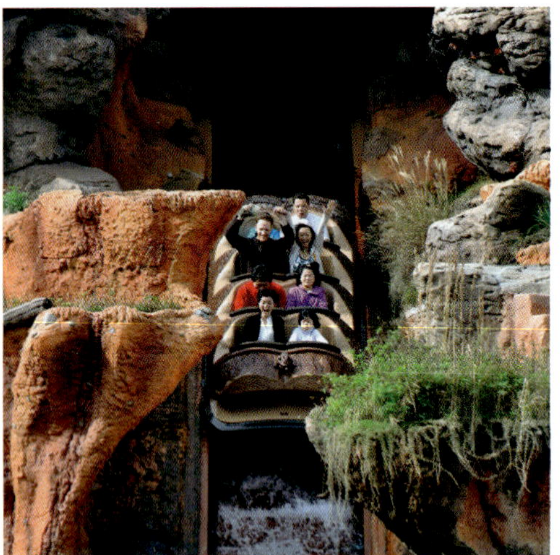

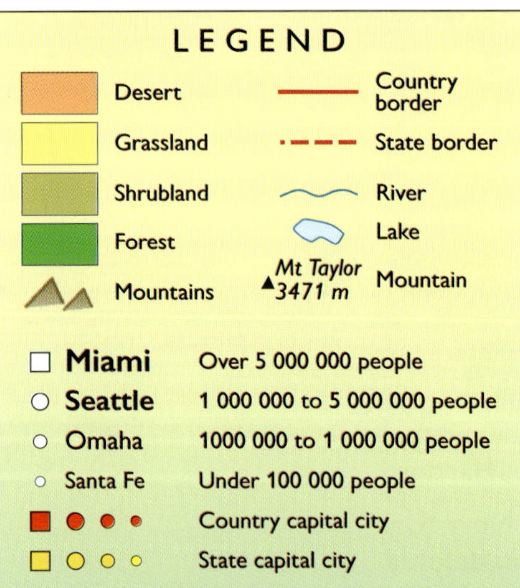

LEGEND

Desert		Country border	
Grassland		State border	
Shrubland		River	
Forest		Lake	
Mountains		Mt Taylor 3471 m	Mountain

Miami	Over 5 000 000 people
Seattle	1 000 000 to 5 000 000 people
Omaha	1000 000 to 1 000 000 people
Santa Fe	Under 100 000 people
	Country capital city
	State capital city

A
120°W
B
110°W
3
40°N
2
30°N
1
20°N
C
Vancouver Island
Vancouver
Victoria
Calgary
Regina
Seattle
Olympia
Mt Rainier 4392 m
Spokane
Mt St Helens 2950 m
Washington
Snake River
Portland
Vancouver
Richland
Salem
Missouri River
Helena
Montana
Eugene
Oregon
Idaho
Yellowstone
Billings
ROCKY MOUNTAINS
Medford
Boise
Grand Teton 4196 m
Cloud Peak 4016 m
Mt Shasta 4317 m
Gannett Peak 4207 m
Redding
Sacramento River
Humboldt River
Great Salt Lake
North Platte River
Wyoming
Reno
Carson City
Salt Lake City
Cheyenne
Santa Rosa
San Francisco
Sacramento
Nevada
Arc Dome 3593 m
Utah
Stockton
San Jose
SIERRA NEVADA
Mt Elbert 4399 m
Denver
Salinas
Delano Peak 3710 m
Colorado Springs
California
Fresno
Mt Whitney 4418 m
Colorado
UNITED
Las Vegas
Grand Canyon
Blanca Peak 4364 m
Santa Barbara
COLORADO PLATEAU
Colorado River
Humphreys Peak 3851 m
Mt Taylor 3471 m
Los Angeles
Long Beach
Anaheim
Santa Fe
Albuquerque
Arizona
San Diego
Phoenix
Baldy Peak 3533 m
New Mexico
Tijuana
Chandler
Mexicali
San Luis
Tucson
Nogales
El Paso
Ciudad Juarez
Rio Grande
Midland
PACIFIC OCEAN
Lower California
Gulf of California
Hermosillo
Point Eugenia
Chihuahua
Delicias
EASTERN
N
W
E
S
Tropic of Cancer
MEXICO
Torreon
Culiacan
Lake Santiaguillo
La Paz
Durango
0
150
300
450 km
1 centimetre on the map measures 135 kilometres on the ground.
Cape San Lucas
Mazatlan
Fresnillo

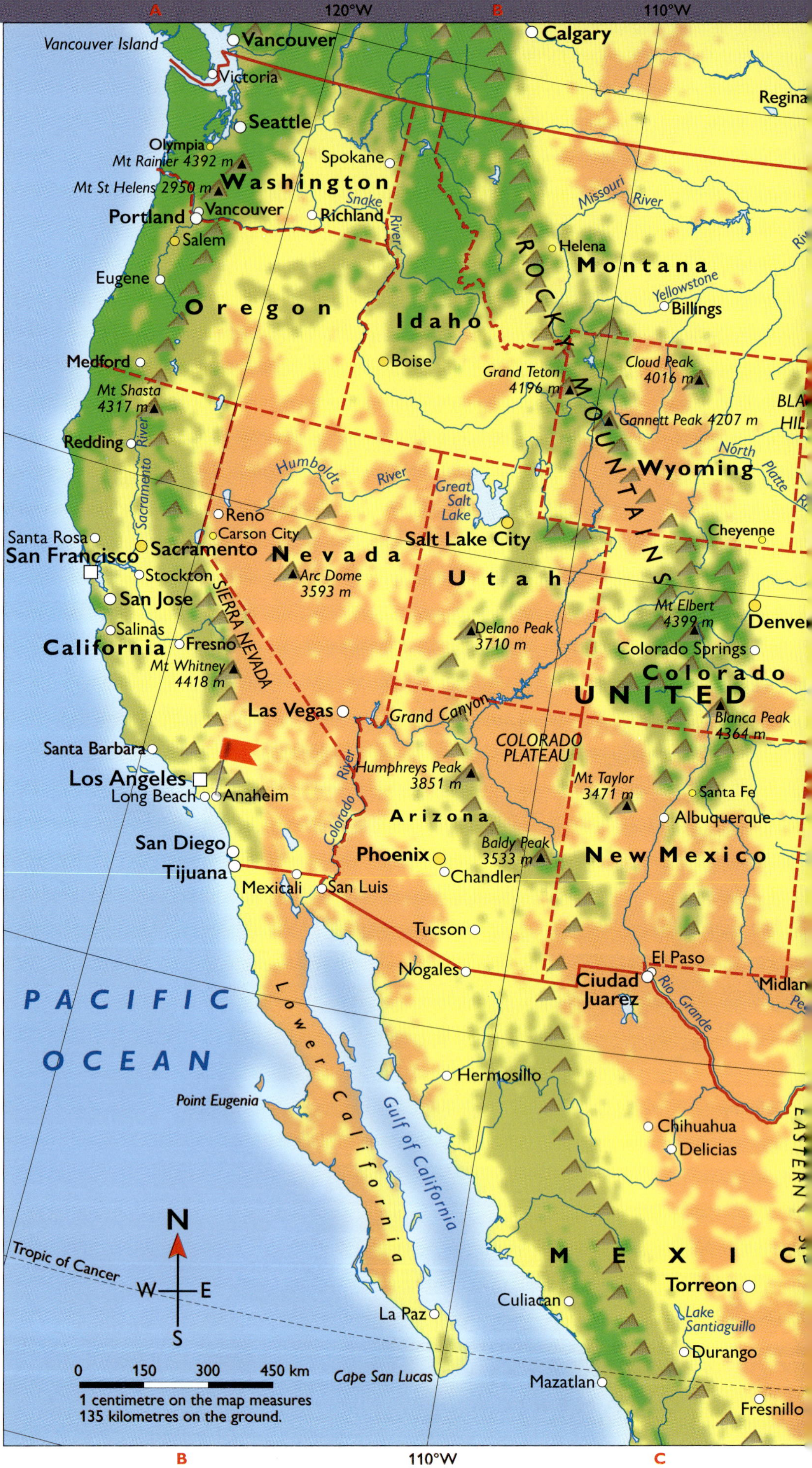

100°W D 90°W E 80°W F 70°W G

CANADA
Lake Winnipeg
Winnipeg
Cochrane
Alma
St Lawrence River
Quebec
Fredericton
Thunder Bay
Lake Superior
Grand Forks
Red River
Trois Rivieres
3
Missouri River
Bismarck
North Dakota
Fargo
Duluth
Michigan
Sudbury
Montreal
Maine
Augusta
Ottawa
Minnesota
Lake Champlain
Vermont
Montpelier
Portland
Georgian Bay
Lake Huron
Kingston
New Hampshire
South Dakota
St Paul
Wisconsin
Lake Michigan
New York
Concord
Manchester
Pierre
Minneapolis
Green Bay
Toronto
Lake Ontario
Massachusetts
Mississippi River
Albany
Syracuse
Boston
Niagara Falls
Rochester
Grand Rapids
Sioux Falls
Milwaukee
London
Buffalo
Hartford
Providence
Rhode Island
Lansing
Connecticut
Madison
Lake Erie
Bridgeport
Waterloo
Detroit
Long Island
40°N
Sioux City
Rockford
Cleveland
New York
Iowa
Chicago
Toledo
Pennsylvania
Nebraska
Trenton
Des Moines
Youngstown
Harrisburg
Philadelphia
Omaha
Fort Wayne
Pittsburgh
New Jersey
Indiana
Ohio
Baltimore
Dover
Lincoln
Bloomington
Columbus
Annapolis
Delaware
Illinois
Indianapolis
Washington DC
Springfield
Cincinnati
West Virginia
Maryland
Kansas City
Bloomington
Kansas
Charleston
Topeka
Jefferson City
Louisville
Huntington
Richmond
St Louis
Frankfort
Lexington-Fayette
APPALACHIAN MOUNTAINS
STATES OF AMERICA
Virginia
Norfolk
Wichita
Missouri
Kentucky
North Carolina
Springfield
Raleigh
Greensboro
Cape Hatteras
2
Nashville
Knoxville
Tulsa
Tennessee
Charlotte
Oklahoma City
Amarillo
Memphis
Greenville
Wilmington
Red River
Oklahoma
Arkansas
South Carolina
Huntsville
Little Rock
Columbia
River
Atlanta
Birmingham
Charleston
Texarkana
Macon
Mississippi
Georgia
Savannah
ATLANTIC OCEAN
30°N
Fort Worth
Dallas
Jackson
Montgomery
Texas
Alabama
Colorado River
Albany
Waco
Mississippi River
Pearl River
Alexandria
Mobile
Jacksonville
Tallahassee
Louisiana
Florida
Austin
Baton Rouge
Panama City
Amistad Reservoir
Houston
Daytona Beach
New Orleans
Apalachee Bay
Pasadena
Orlando
San Antonio
Nueces River
Tampa
Piedras Negras
Bradenton
Lake Okeechobee
Rio Grande
Corpus Christi
1
Monclova
Fort Lauderdale
Naples
Miami
McAllen
Nassau
Gulf of Mexico
Brownsville
Cape Sable
Monterrey
Florida Keys
BAHAMAS
Long Island
Ciudad Victoria
Havana
Matanzas
Pinar del Rio
CUBA
Camaguey

100°W D 90°W E 80°W F

Central America and the Caribbean

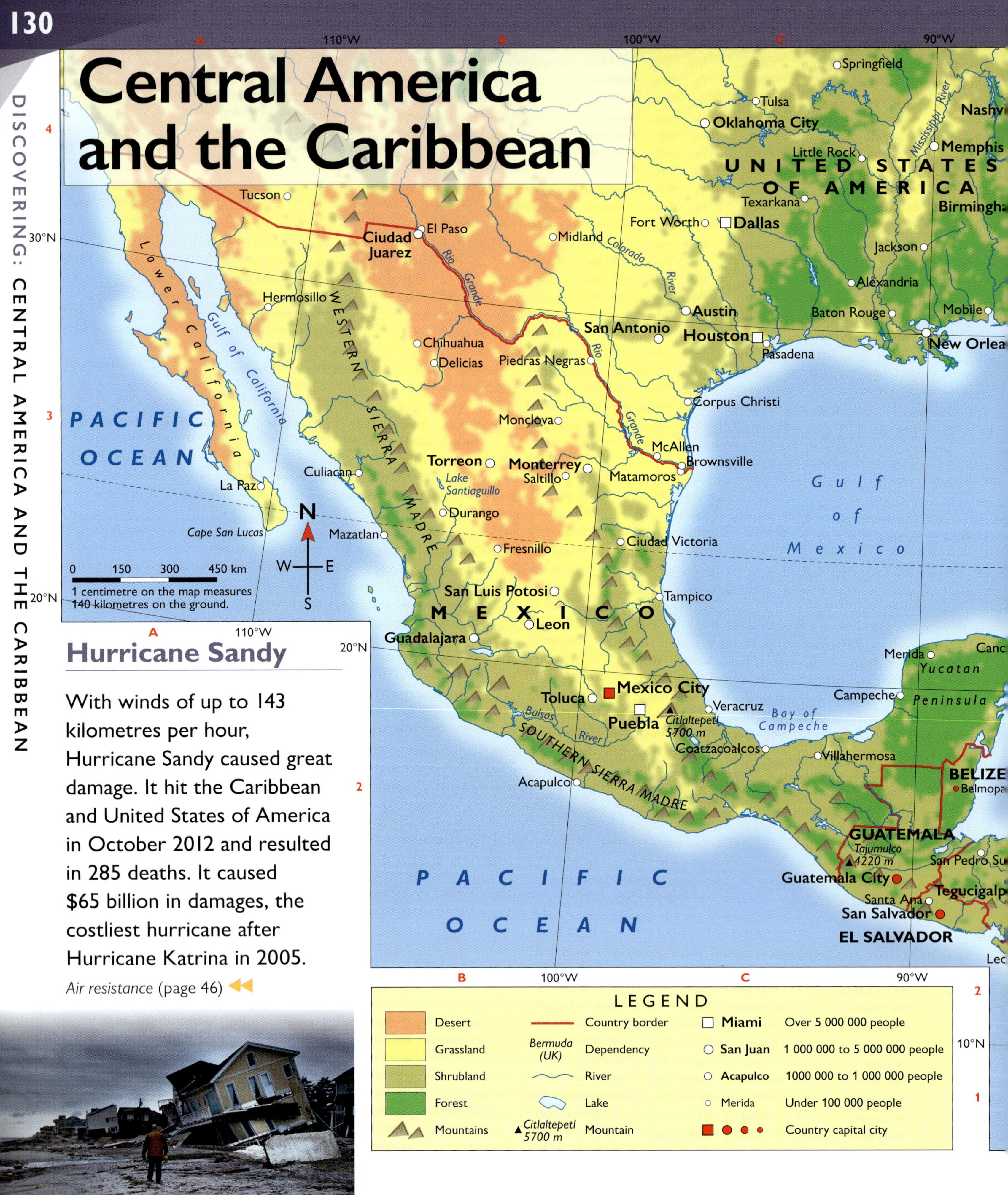

Hurricane Sandy

With winds of up to 143 kilometres per hour, Hurricane Sandy caused great damage. It hit the Caribbean and United States of America in October 2012 and resulted in 285 deaths. It caused $65 billion in damages, the costliest hurricane after Hurricane Katrina in 2005.

Air resistance (page 46)

ATLANTIC OCEAN
CARIBBEAN SEA
Richmond
Norfolk
Knoxville
Mt Mitchell 2037 m
Greensboro
Raleigh
Cape Hatteras
Charlotte
Huntsville
Greenville
Savannah River
Columbia
Atlanta
Bermuda (UK)
Macon
Charleston
Montgomery
Savannah
Albany
Tallahassee
Panama City
Jacksonville
Apalachee Bay
Daytona Beach
Orlando
Tampa
Bradenton
Grand Bahama
Naples
Fort Lauderdale
Miami
Eleuthera
Nassau
BAHAMAS
Tropic of Cancer
Cape Sable
Andros
New Providence
Florida Keys
Florida Strait
South Andros
CUBA
Havana
Crooked Island
Caicos Islands (UK)
Cienfuegos
Turks Islands (UK)
Bay of Pigs
Great Inagua
DOMINICAN REPUBLIC
Guantanamo
Cayman Islands (UK)
Santiago de Cuba
Duarte Peak 3174 m
Virgin Islands (UK)
Anguilla (UK)
St Martin (France/Netherlands)
Mona Passage
San Juan
ANTIGUA and BARBUDA
HAITI
Grand Cayman
Port-au-Prince
Santo Domingo
Basseterre
St John's
Montego Bay
Puerto Rico (USA)
ST KITTS and NEVIS
Montserrat (UK)
Kingston
JAMAICA
Hispaniola
Leeward Islands
Guadeloupe (France)
DOMINICA
Roseau
Martinique (France)
ST LUCIA
Castries
HONDURAS
Windward Islands
BARBADOS
Kingstown
Bridgetown
Coco River
ST VINCENT and the GRENADINES
Puerto Cabezas
GRENADA
St George's
NICARAGUA
Guajira Peninsula
Aruba (Netherlands)
TRINIDAD and TOBAGO
Lake Managua
San Andres (Colombia)
Coro
Port of Spain
Managua
Bluefields
Cristobal Colon 5797 m
Caracas
Trinidad
Lake Nicaragua
Barranquilla
Maracaibo
Valencia
Maracay
Barcelona
Cartagena
PERIJA MOUNTAINS
Lake Maracaibo
Barquisimeto
San Jose
Panama Canal
Valera
Ciudad Guayana
Chirripo Grande 3819 m
Gulf of Darien
Ciudad Bolivar
COSTA RICA
Panama City
Bolivar Peak 5007 m
Orinoco River
La Palma
PANAMA
Cucuta
GUYANA
Gulf of Panama
COLOMBIA
VENEZUELA
Bucaramanga
Angel Falls
80°W
70°W
60°W
30°N
20°N
10°N
D
E
F
G
1
2
3
4

South America

South America is divided by the Andes Mountains. To the west is desert and to the east is rainforest. The Amazon rainforest is the largest rainforest in the world. Brazil is the largest country in South America. Half of the continent's people live there. Christianity, in particular Roman Catholicism, is the most popular religion.

People of the Amazon rainforest

1

The Xingu Indians live in the Amazon rainforest in Brazil. They are skillful hunters. The Xingu Indians hunt for monkeys, wild pigs, fish and birds.

Australian Aboriginal people and hunting (page 53)

South America facts

Population:	420 000 000
Largest city:	Sao Paulo 21 300 000 people

Amazon River

2

The Amazon River carries more water than any other river on Earth. It is estimated that one-sixth of all the fresh water that drains into the world's oceans is from the Amazon River.

Rivers change the land (page 22)

LEGEND

(red line)	Country border
BRAZIL	Country name
Aruba (Netherlands)	Dependency

Country capital city

■ **Santiago**	Over 5 000 000 people
● **Brasilia**	1 000 000 to 5 000 000 people
● Sucre	100 000 to 1 000 000 people
• (no example)	Under 100 000 people

Giant anaconda

Anacondas live in the Amazon rainforest. They are among the world's largest snakes. Anacondas grow up to six metres long. Adult anacondas can eat sheep, dogs and even jaguars.

Feeding (page 34)

Pope Francis I

On 13 March 2013, Pope Francis I was elected as head of the Roman Catholic Church. Born in Buenos Aires, the capital city of Argentina, he is the first South American pope.

Different religions (page 67)

Northern South America

The Uros

The indigenous people of Lake Titicaca, the Uros, live on man-made floating islands. These islands are made of dried bundles of totora reeds. The homes and boats of the Uros are also made from totora reeds.

People at home (page 63)

Large islands house up to ten families.

An animal head decorates a traditional reed boat.

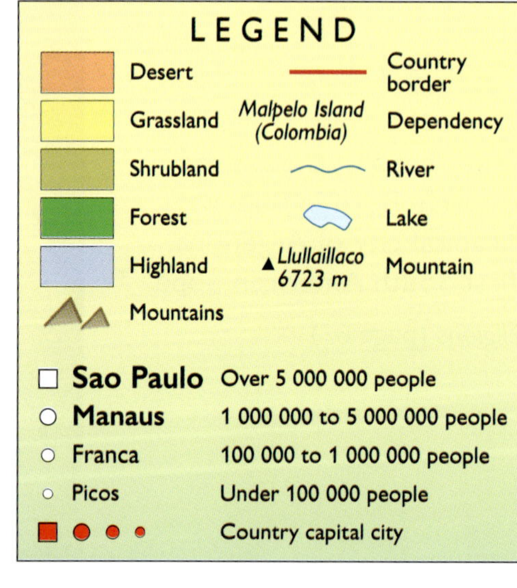

Martinique (France)
ST LUCIA
ST VINCENT and the GRENADINES
BARBADOS
GRENADA
Margarita
Caracas
Maracay
Barcelona
Port of Spain
Trinidad
TRINIDAD and TOBAGO
ATLANTIC OCEAN
Orinoco River
Ciudad Guayana
Ciudad Bolivar
VENEZUELA
GUYANA
Georgetown
Paramaribo
Cayenne
Puerto Paez
Angel Falls
Roraima 2810 m
GUIANA HIGHLANDS
SURINAME
French Guiana
Saul
Maroa
Boa Vista
Branco River
Uaupes
Negro River
Barcelos
Macapa
Mouths of the Amazon
Marajo Island
Cape Maguarinho
Equator
Belem
Sao Luis
Parnaiba
Santarem
Manaus
Amazon River
Sobral
Fortaleza
Caxias
Teresina
Cape Sao Roque
Selvas
Manicore
Natal
Picos
Joao Pessoa
Campina Grande
Gradaus
Recife
Madeira River
Porto Velho
Petrolina
Rio Branco
Maceio
Palmas
Riberalta
Juruena River
Guapore River
BRAZIL
Barreiras
Feira de Santana
Sao Francisco River
Salvador
Apolo
Lake Titicaca
BOLIVIA
PLATEAU OF MATO GROSSO
Cuiaba
Caceres
Ilheus
Vitoria da Conquista
BRAZILIAN HIGHLANDS
Ancohuma 6388 m
La Paz
Brasilia
Goiania
Montes Claros
Cochabamba
Santa Cruz
Rio Verde
Parnaiba River
Teofilo Otoni
Sucre
Potosi
Uberlandia
Sete Lagoas
Belo Horizonte
Campo Grande
Franca
Vitoria
Aucanquilcha 6180 m
Tarija
Ribeirao Preto
Calama
Puerto Pinasco
Dourados
Agulhas Negras 2786 m
Pilcomayo River
PARAGUAY
Campinas
Londrina
Rio de Janeiro
Sao Paulo
Santos
Tropic of Capricorn
Salta
Llullaillaco 6723m
Asuncion
Foz do Iguacu
ARGENTINA
Curitiba
Formosa
60°W
50°W
40°W
10°N
0°
10°S
20°S
C
D
E
F
5
4
3
2
1

Southern South America

World's driest place

The Atacama Desert in Chile is the driest place in the world. Some parts of the Atacama Desert have not had rain for over 400 years. Plants and animals survive by taking moisture from fog. The fog comes in from the cold Pacific Ocean nearby. Cactus plants survive in the Atacama Desert by catching the water in the fog on their spikes. The water then drops to the ground where it is taken in by the roots of the cactus.

World's hottest place (page 40)

Drought in the Atacama Desert transforms a dry creek bed into cracked mud.

This type of cactus can survive in the dry Atacama Desert.

BOLIVIA
PARAGUAY
BRAZIL
URUGUAY
ARGENTINA
CHILE
PACIFIC OCEAN
ATLANTIC OCEAN
ANDES MOUNTAINS
Atacama Desert
PATAGONIAN DESERT
Pampas
Tropic of Capricorn
Sucre
Potosi
Iquique
Aucanquilcha 6180 m
Tarija
Calama
Antofagasta
Pular 6225 m
Salta
Copiapo
Incahuasi 6620 m
San Miguel de Tucuman
Santiago del Estero
La Rioja
Salinas Grandes
Coquimbo
San Juan
Cordoba
Santa Fe
Rosario
San Nicolas
Mt Aconcagua 6962 m
Mendoza
Valparaiso
Santiago
Mt Tupungato 6800 m
Maipo 5322 m
Buenos Aires
La Plata
River Plate
Montevideo
Curico
Bolivar
Tandil
Mar del Plata
Concepcion
Los Angeles
Colorado River
Bahia Blanca
Temuco
Neuquen
Viedma
Gulf of San Matias
Valdes Peninsula
Tronador 3536 m
Puerto Montt
Chiloe Island
Mt Chato 2420 m
Trelew
Mt Alto Nevado 2256 m
Chonos Archipelago
Comodoro Rivadavia
Peak San Valentin 4058 m
Puerto Deseado
San Lorenzo 3699 m
Mt Fitzroy 3374 m
Upsala Glacier
Wellington Island
Santa Cruz
Gallegos
Falkland Islands (UK)
West Falkland
Stanley
East Falkland
Falkland Sound
Punta Arenas
Tierra del Fuego
Ushuaia
Mt Darwin 2438 m
Cape San Diego
Staten Island
Cape Horn
Drake Passage
South Georgia (UK)
South Sandwich Islands (UK)
Fort Olimpo
Puerto Pinasco
Dourados
Campo Grande
Asuncion
Formosa
Corrientes
Salado
Parana
Uruguay
River
Guarapuava
Foz do Iguacu
Londrina
Marilia
Franca
Patos
Grande River
Belo Horizonte
Vitoria
Campinas
Sorocaba
Sao Paulo
Santos
Rio de Janeiro
Curitiba
Joinvile
Florianopolis
Passo Fundo
Santa Maria
Porto Alegre
Concordia
Pelotas
Patos Lagoon
Rio Grande
Lake Mirim
N
W
E
S
0 200 400 600 km
1 centimetre on the map measures 185 kilometres on the ground.
20°S
30°S
40°S
50°S
70°W
60°W
50°W
80°W
40°W
30°W
A B C D E F G
1 2 3 4 5

Antarctica

Antarctica is the coldest, windiest and driest continent. Covering an area of 14 000 000 square kilometres, it is nearly twice the size of Australia. In winter, Antarctica doubles in size due to the sea ice that forms around the coast. Approximately 99 per cent of Antarctica is covered by ice. This ice stores around 80 per cent of the world's fresh water. There is no permanent human population, although 4000 scientists and researchers visit Antarctica every year.

Antarctica facts
Population: 4000 (not permanent)
Highest point: Vinson Massif 5140 m

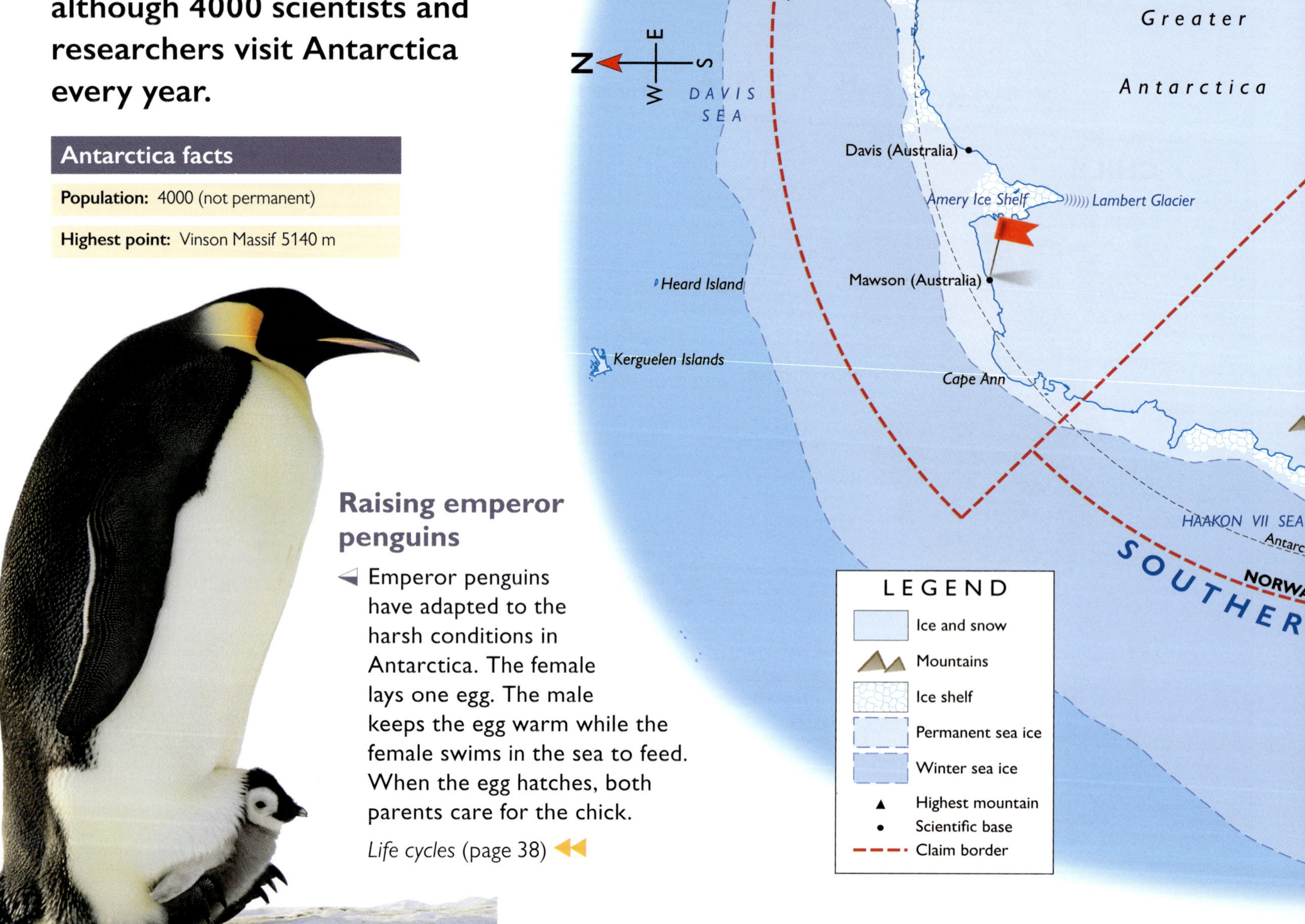

Raising emperor penguins

◄ Emperor penguins have adapted to the harsh conditions in Antarctica. The female lays one egg. The male keeps the egg warm while the female swims in the sea to feed. When the egg hatches, both parents care for the chick.

Life cycles (page 38) ◄◄

Aurora australis

Aurora australis or the southern lights appears in Antarctic skies in winter. Amazing colours and shapes are produced by streams of charged particles from the Sun (solar wind) crashing into Earth's atmosphere.

Earth, Sun and Moon (page 16)

Mawson Station

Mawson Station is Australia's oldest scientific station in Antarctica. In summer, it becomes a giant outdoor science laboratory where scientists study everything from seals to climate change.

Sir Douglas Mawson (page 57)

Entry **Page** **Grid reference**

Entry **Page** **Grid reference**

C

D

Entry **Page** **Grid reference**

H

Entry **Page** **Grid reference**

L

Entry **Page** **Grid reference**

N

Entry **Page** **Grid reference**

O

P

Q

R

Entry **Page** **Grid reference**

T

Entry **Page** **Grid reference**

W

X

Y

Z

ACKNOWLEDGEMENTS

The author and the publisher wish to thank the following copyright holders for reproduction of their material.

Imagefolk/Gerard Lacz, front and back cover.

Photos: AAP Image/AP Photo/Vincent Yu, p. 67 (dragon)/Lukas Coch, 65 top/Peter Eve, 54 left/Richard Wainwright, 64 bottom/Tracey Nearmy, 68-69 (parade); **Age Fotostock**/Eric Lafforgue, 63 Africa/H J Igelmund, 119 (giraffe)/Morales, 133 (anaconda)/S. T. Yiap, contents (opera), 70–71 (opera house)/Ton Koene, contents (africa), 90-91/travelstock44, 98 (haka); **Alamy**/ GIPhotoStock/Visuals Unlimited, 51 left/Andrew Bain, 30 (bushwalking)/blickwinkel/McPHOTO/REB, 72 (emu)/Claire Leimbach/Robert Harding World Imagery, 52 left/Cultura Creative, 34 (breath)/David Ball, 104 (robots)/David Hancock, 62 Australia/David Wall, 1, 24 right, 73 (Sea world), contents: (balloon)/Devy Masselink, 111 (frozen canal)/Ettore Ferrari/epa, 133 (Pope)/Friedrich Stark, 63 (Singapore)/Gaertner, 63 (native American)/Genevieve Vallee, 55 top/Hemis, 136 (cactus)/J Garden, 35 (fly trap)/JLImages, 46 bottom/John Sylvester, 125 left and right/Megapress, 62 (Canada)/ nobleIMAGES, 62 (India)/Nucleus Medical Art Inc, 38 (foetus)/Photoshot, 39 (butterfly eggs)/Red Poppy, 128/Robert Harding World Imagery, 84 (roll of honour)/ Skyscan, 63 (England)/Sylvain Grandadam/Robert Harding World Imagery, 53 bottom/Thant Zaw Wai, 84 (War Memorial)/Ton Koene/Visuals Unlimited, 132 (Xingu boy)/Yves Marcoux/First Light, 124 (L'Anse aux Meadows); **Ardea**/Jean-Marc La Roque, 22–23 (background); **Australian Antarctic Division**/Chris Wilson, 139 (Aurora)/Gary Dowse, 139 (Mawson station); **Australian War Memorial**, H03500, 114 (Anzac Cove), REL32862.001, 69 (hat); **Centro**/Victoria Gardens, 12–13 bottom; **Chief Minister Support and Protocol, Canberra**, 84 (coat of arms), 84 (flag); **City2See**, 2; Courtesy of **December Media** Pty Ltd, photographer Mike Baker, 86 (William Buckley); Permission to reproduce the Commonwealth Coat of Arms granted by the **Department of the Prime Minister and Cabinet**, 72 (coat of arms), 72 (Flag); **Department of Attorney General and Justice, New South Wales**, 82 (coat of arms), 82 (flag); **Department of Premier and Cabinet, Victoria**, 80 (sea dragon), 86 (gold), 86 (honeyeater), 86 (pink heath), 86 (possum), 86 (coat of arms), 86 (flag); **Department of the Chief Minister, Northern Territory**, 76 (coat of arms), 76 (flag); **Department of Premier and Cabinet, South Australia**, 78 (coat of arms), 78 (flag); **Department of Premier and Cabinet, Queensland**, 80 (coat of arms), 80 (flag); **Department of Premier and Cabinet, Western Australia**, 74 (coat of arms), 74 (flag); **Department of Premier and Cabinet, Tasmania**, 88 (coat of arms), 88 (flag); **Dreamstime**/Jeecis, 19 centre; **Fairfax Media**/Brendan Esposito, 67 (Ramadan) /Chris Lane, 60 bottom/Glenn Campbell, 55 bottom/Paul Rovere, 30, 32 (logging)/Peter Mathew, 88 (centurion)/Peter Rae, 61 bottom/Robert Rough, 30 (coal); **Fotolia**/Ingo Bartussek, 48 (slipping); **Getty Images**/Breckeni, 36 (tyrannosaurus)/ Blaine Harrington III, 20/iStock, 32 (oil rig) /AFP, 74 (Burrup rock carvings)/ AFP Photo/William West, 61 top/Alfred Pasieka, 50 bottom/Anders Ryman, 98 (conch)/Auscape/UIG, 38 (crocodiles)/Bloomberg, 33 (plastic mould)/Christopher Groenhout, 73 (Sydney)/Colin Keates, 36–37 (Diprotodon)/Daniel H. Bailey, contents (mountain), 14–15/Danita Dellmont, 101 (panda)/Danny Lehman, 63 (PNG)/ David Wall Photo, 45 (lizard)/DLILLC, 138 (penguin)/Gallo Images, 120 (chimpanzee)/Gavin Hellier, 134 (reed boat)/Grant Faint, 124 (Grand Canyon)/Greg Baker, 49 top/Imageplus, 63 (Brazil)/In Pictures/Barry Lewis, 74 (Tasmanian tiger)/John and Lisa Merrill, 45 (glass blowing)/John Short/Design Pics, 110 (erosion)/John W Banagan, 31 (boat)/Jose Fuste Raga, 108 (Burj Khalifa)/Kevin Schafer/Minden Pictures, 118 (Silky sifaka)/Layne Kennedy, 132 (Amazon River)/Martin Harvey, 106 (Ganges)/mevans, 49 bottom/Michael Freeman, 101 (mine)/Michael Nichols, 34 (tree)/Michael S. Lewis, 119 (gold mining)/Minden Pictures/Luciano Candisani, 29 bottom/Minden Pictures/Tim Fitzharris/, 40 (death valley)/Paul A. Souders, 63 (snowmobile) /Paul Chesley, 125 (snake)/Peter Walton Photography, 31 (crop)/Rainer Schlegelmilch, 42-43 (racing car)/Sally Anscombe, 19 top/Science Picture Co/Science Faction, 116 (mammoth)/Scientifica, 43 (thermogram)/Scott Thistlethwaite, 30 (wind)/Sean Justice, 50 top/Simon Marcus, 42 (family)/Theo Allofs, 136 (desert)/Tom Bonaventure, 8 top left/VCG, 46 top/Vincent J. Musi/National Geographic Society, 62 (Turkey)/Wayne Lynch, 52 centre/Spencer Platt, 130 (hurricane); **Imagefolk**/Jonathan and Angela Scott, 35 (cheetah)/Minden Pictures, back cover (right)/Nature Picture Library/ Owen Newman, 88 (nest)/Nature Picture Library/Jane Burton, 35 (chick); **Imagine China**/Wang Qingxiang, 100 (temple); **Ingram publishing**, 12–13; **istockphoto**/Bubaone, 40 (thermometer)/ImageGap, 48 (smooth tyre)/Jack Puccio, 32 (woodchips)/JazzIRT, 44 (oven)/Jerome Skiba, 48 (mountain-bike tyre)/Quan Long, 32 (blast furnace)/Rob Broek, 112 (blue lagoon); **Lavinia Nixon and Channel Nine**, 13 bottom; **Lindsay Edwards**, 66; Megafauna stamp image by **Peter Trusler**, 2008; © Australian Postal Corporation, 36-37 (megafauna); **Mitchell Library**, State Library of NSW – ML 1222, 58 Arthur Phillip; Moore-Jones, Horace Millichamp, 1867? –1922. Moore-Jones, Horace Millichamp 1878–1922: To the memory of our hero comrade 'Murphy' (Simpson) killed May 1915. Heroes of the Red Cross. Private Simpson, D.C.M., & his donkey at Anzac. Printed in England and published by W. J. Bryce, 24a Regent Street, London, S.W. 1. 1918. Ref: C-057-002. **Alexander Turnbull Library**, Wellington, New Zealand. http://beta.natlib.govt.nz/records/22325152, p. 69 (Simpson); **Harold Thomas**, 60 top left; **Murray River Photos**, 52 right; **NASA**, 17 top left, 17 top right, 47 (astronaut)/SDO (AIA), 16; **National Library Australia**, Jukes, Francis, 1746–1812, nla.pic-an6016289, 82 (Sydney cove)/Broad, Alf Scott, 1854–1929, nla.pic-an8955120, 58 (hunting sheep); **Newspix**/Brett Faulkner, 76 (Jim Jim Falls)/ Ian Munro, 32 (iron ore mine)/Jon Hargest, 65 bottom/Mark Calleja, 22 bottom/Mark Cranitch, 80 mine/Shane Bell, 43 (children running)/Wayne Ludbey, 62 (dawn service); **Oceanwide Images**/Gary Bell, 10, 29 top, back cover (left); **One Shot**/PNZ/Ralph Talmont, 98 (hongi); **Oxford Universtiy Press**/Michelle Shipp, 69 (Anzac pin); **Peter Van Noorden**, 18–19 top, 19 bottom; **Science Photo Library**/ Ria Novosti, 40 (Vostok), 40–41 (background)/Jaime Chirinos, 78 (marsupial lion); **Shutterstock**, 102 (Halong Bay), 111 (Eiffel Tower), 134 (floating islands), 20–21 bottom, 20-21 top, 20–21 top, 21, 21 (calendar), 22 top, 24 left, 24-25 (background), 30 (koala), 32 (iron ore), 32 (oil), 32 (oil refinery), 32-33 (paper background), 32–33 (plastic background), 32-33 (steel background), 33 (bike), 33 (car), 33 (duck), 33 (flaming furnace), 33 (girl), 33 (molten steel), 33 (resin pellets), 34 (boy), 35 (stork), 38 (leaves), 39 (transforming to pupa), 39 (butterfly), 39 (caterpillar), 39 (transforming to butterfly), 4 (female), 4 (man), 42 (toaster), 44 (feet), 44 (heater), 45 (chocolate), 45 (ice), 47 (bouncing ball), 48–49 (background), 51 right, 53 top, 54 right, 54–55, 58–59 (background), 59 (portrait frame), 60–61, 62–63 (background), 63 (bullet train), 64 top, 64–65, 66 (noodles), 66 (pizza), 66 (tortillas), 67 (envelope), 68–69 (poppy field), 69 (poppy), 78 (skeleton), 8 bottom right, 8 bottom right, 8 top right, 80 (coal); **Taronga Western Plains Zoo**, 4–5/Rick Stevens, 4 (giraffes); **The Natural History Museum**, London, 59 (Bennelong), 59 (boat scene); **Torres Strait Island Regional Council**/Mr Bernard Namok, 60 top right.

Every effort has been made to trace the original source of copyright material contained in this book. The publisher will be pleased to hear from copyright holders to rectify any errors or omissions.

OXFORD
UNIVERSITY PRESS

Oxford University Press is a department of the University of Oxford. It furthers the University's objective of excellence in research, scholarship, and education by publishing worldwide. Oxford is a registered trademark of Oxford University Press in the UK and in certain other countries.

Published in Australia by
Oxford University Press
Level 8, 737 Bourke Street, Docklands, Victoria 3008, Australia

© Oxford University Press 2018

The moral rights of the author have been asserted

First published 2013
Revised edition 2018
Reprinted 2018, 2019, 2020, 2021, 2022, 2023 (twice), 2025 (twice)

All rights reserved. No part of this publication may be reproduced, stored in a retrieval system, or transmitted, in any form or by any means, without the prior permission in writing of Oxford University Press, or as expressly permitted by law, by licence, or under terms agreed with the reprographics rights organisation. Enquiries concerning reproduction outside the scope of the above should be sent to the Rights Department, Oxford University Press, at the address above.

You must not circulate this work in any other form and you must impose this same condition on any acquirer.

National Library of Australia Cataloguing-in-Publication data
Title: Oxford atlas+ for Australian schools. 3-4 / Oxford University Press.
Edition: Revised edition.
ISBN: 978 0 19 031077 6 (paperback)
Notes: Includes index.
Target Audience: For primary school age.
Subjects: Atlases, Australian–Juvenile literature.

Reproduction and communication for educational purposes
The Australian *Copyright Act 1968* (the Act) allows a maximum of one chapter or 10% of the pages of this work, whichever is the greater, to be reproduced and/or communicated by any educational institution for its educational purposes provided that the educational institution (or the body that administers it) has given a remuneration notice to Copyright Agency Limited (CAL) under the Act.

For details of the CAL licence for educational institutions contact:

Copyright Agency Limited
Level 15, 233 Castlereagh Street
Sydney NSW 2000
Telephone: (02) 9394 7600
Facsimile: (02) 9394 7601
Email: info@copyright.com.au

'Indigenous sustainability practices' (pp. 54–55), 'Living cultures' (pp. 60–61) and 'Democracy' (pp. 64–65) written by Rachel Kennedy
Project management by Katrina Spencer
Designed and typeset by Rose Keevins and Oxford University Press
Illustrated by Robert Mancini, Alan Laver and Guy Holt
Cartography and map index by MAPgraphics Pty Ltd, Brisbane
Subject index by Max McMaster
Printed in Hong Kong by Sheck Wah Tong Printing Press Ltd.

Links to third party websites are provided by Oxford in good faith and for information only. Oxford disclaims any responsibility for the materials contained in any third party website referenced in this work.

Aboriginal and Torres Strait Islander people are warned that this book may contain images of deceased people.

AUSTRALIA AND THE PACIFIC
Samoa
Brunei
Kyrgyzstan
Singapore
EUROPE
France
Monaco
Solomon Islands
Cambodia
Laos
South Korea
Germany
Montenegro
Australia
Tonga
China
Lebanon
Sri Lanka
Albania
Greece
Netherlands
Cook Islands
Tuvalu
Cyprus
Malaysia
Syria
Andorra
Hungary
North Macedonia
Federated States of Micronesia
Vanuatu
Georgia
Maldives
Taiwan
Austria
Iceland
Norway
Fiji
India
Mongolia
Tajikistan
Belarus
Ireland
Poland
Kiribati
ASIA
Indonesia
Myanmar
Thailand
Belgium
Italy
Portugal
Marshall Islands
Iran
Nepal
Timor-Leste
Bosnia and Herzegovina
Kosovo
Romania
Nauru
Afghanistan
Iraq
North Korea
Turkey
Bulgaria
Latvia
Russia
New Zealand
Armenia
Israel
Oman
Turkmenistan
Croatia
Liechtenstein
San Marino
Niue
Azerbaijan
Japan
Pakistan
United Arab Emirates
Czech Republic
Lithuania
Serbia
Northern Marianas
Bahrain
Jordan
Philippines
Uzbekistan
Denmark
Luxembourg
Slovakia
Palau
Bangladesh
Kazakhstan
Qatar
Vietnam
Estonia
Malta
Slovenia
Papua New Guinea
Bhutan
Kuwait
Saudi Arabia
Yemen
Finland
Moldova
Spain